MOLECULAR
BIOLOGY
INTELLIGENCE
UNIT

CYTOKINES PRODUCED BY POLYMORPHONUCLEAR NEUTROPHILS: MOLECULAR AND BIOLOGICAL ASPECTS

Marco A. Cassatella, M.D.
Institute of General Pathology
Verona, Italy

Springer
New York Berlin Heidelberg London Paris
Tokyo Hong Kong Barcelona Budapest

R.G. LANDES COMPANY
AUSTIN

MOLECULAR BIOLOGY INTELLIGENCE UNIT
CYTOKINES PRODUCED BY POLYMORPHONUCLEAR NEUTROPHILS: MOLECULAR AND BIOLOGICAL ASPECTS

R.G. LANDES COMPANY
Austin, Texas, U.S.A.

International Copyright © 1996 Springer-Verlag, Heidelberg, Germany

All rights reserved.
No part of this book may be reproduced or transmitted in any form or by any means, electronic or mechanical, including photocopy, recording, or any information storage and retrieval system, without permission in writing from the publisher.
Printed in the U.S.A.

Please address all inquiries to the Publishers:
R.G. Landes Company, 909 Pine Street, Georgetown, Texas, U.S.A. 78626
Phone: 512/ 863 7762; FAX: 512/ 863 0081

International distributor (except North America):
Springer-Verlag GmbH & Co. KG
Tiergartenstrasse 17, D-69121 Heidelberg, Germany

 Springer

International ISBN: 3-540-61418-4

While the authors, editors and publisher believe that drug selection and dosage and the specifications and usage of equipment and devices, as set forth in this book, are in accord with current recommendations and practice at the time of publication, they make no warranty, expressed or implied, with respect to material described in this book. In view of the ongoing research, equipment development, changes in governmental regulations and the rapid accumulation of information relating to the biomedical sciences, the reader is urged to carefully review and evaluate the information provided herein.

Library of Congress Cataloging-in-Publication Data

Cassatella, Marco A.
  Cytokines produced by polymorphonuclear neutrophils: molecular and biological aspects / Marco A. Cassatella.
    p. cm. — (Molecular biology intelligence unit)
  Includes bibliographical references and index.
  ISBN 1-57059-378-7 (alk. paper) : ISBN 0-412-10991-3 (alk. paper)
  1. Neutrophils. 2. Cytokines.   I. Title. II. Series.
QR185.8.N47C37 1996
616.07'9—dc20

                                          95-23860
                                              CIP

Publisher's Note

R.G. Landes Company publishes six book series: *Medical Intelligence Unit, Molecular Biology Intelligence Unit, Neuroscience Intelligence Unit, Tissue Engineering Intelligence Unit, Biotechnology Intelligence Unit and Environmental Intelligence Unit.* The authors of our books are acknowledged leaders in their fields and the topics are unique. Almost without exception, no other similar books exist on these topics.

Our goal is to publish books in important and rapidly changing areas of bioscience and environment for sophisticated researchers and clinicians. To achieve this goal, we have accelerated our publishing program to conform to the fast pace in which information grows in bioscience. Most of our books are published within 90 to 120 days of receipt of the manuscript. We would like to thank our readers for their continuing interest and welcome any comments or suggestions they may have for future books.

Deborah Muir Molsberry
Publications Director
R.G. Landes Company

CONTENTS

1. **The Neutrophil** 1
   - Introduction 1
   - Neutrophil Differentiation and Maturation 1
   - Neutrophil Receptors 2
   - Neutrophil Microbicidal Mechanisms 4
   - The Role of Neutrophils in Acute Inflammation 5
   - Neutrophils in Human Diseases 7

2. **The Cytokines** 9
   - A Definition of Cytokine 9
   - Interferon-α (IFNα) 11
   - Interferon-γ (IFNγ) 12
   - Transforming Growth Factor-β (TGFβ) 13
   - Tumor Necrosis Factor-α (TNFα) 14
   - The Interleukin-1 System (IL-1α/β and IL-1ra) 15
   - Interleukin-2 (IL-2) 17
   - Interleukin-4 (IL-4) and IL-13 17
   - Interleukin-6 (IL-6) 18
   - Interleukin-8 (IL-8) 19
   - Interleukin-10 (IL-10) 21
   - Interleukin-12 (IL-12) 22
   - Growth Related Gene Product-α (GROα)/Melanoma Growth-Stimulatory Activity (MGSA) 23
   - Macrophage Inflammatory Protein-1α and -1β (MIP-1α/β) 23
   - Monocyte Chemotactic Proteins (MCP-1, MCP-2 and MCP-3) 24
   - Interferon Inducible Protein-10 (IP-10) 24
   - Stem Cell Factor (SCF) 25
   - Colony Stimulating Factors (CSF) 25

3. **Main Characteristics of Cytokine Production by Human Neutrophils** 39
   - In Vitro Production of Cytokines by Human Neutrophils: General Features 39
   - Stimuli that Induce Cytokine Production by Neutrophils 41
   - Cytokine Gene Expression 45
   - Cytokine Networks Regulating Neutrophil-Derived Cytokines 49
   - Cytokine Production by Neutrophils In Vivo 51

4. **Production of Specific Cytokines by Neutrophils In Vitro** 59
   Interferon-α (IFNα) 59
   Transforming Growth Factor-β (TGFβ) 61
   Tumor Necrosis Factor-α (TNFα) 62
   Interleukin-1α and -β (IL-1α and -β) 66
   Interleukin-1 Receptor Antagonist (IL-1ra) 70
   Interleukin-6 (IL-6) 77
   Interleukin-8 (IL-8) 80
   Interleukin-12 (IL-12) 87
   Growth Related Gene Product-α (GROα) and GROβ 91
   Macrophage Inflammatory Protein-1α and -1β (MIP-1α/β) ... 94
   Monocyte Chemotactic Proteins (MCP-1, MCP-2 and MCP-3) 96
   Interferon Inducible Protein-10 (IP-10) 97
   CD30 Ligand (CD30L) 98
   Other Cytokines 98

5. **Modulation of Cytokine Production in Human Neutrophils** 113
   Effects of Interleukin-4 (IL-4) 114
   Effects of Interleukin-10 (IL-10) 117
   Effects of Interferon-γ (IFNγ) 125
   Effects of Inflammatory Microcrystals 128
   Effects of Other Substances 129

6. **Molecular Regulation of Cytokine Production in Neutrophils** 135
   Effects of Metabolic Inhibitors 136
   Transcriptional and Post-Transcriptional Regulation 139
   Translational Regulation 146
   Post-Translational Regulation 148
   Additional Remarks 149

7. **Production of Cytokines by Neutrophils Isolated from Individuals Affected by Different Human Pathologies** ... 155
   Production of Interleukin-1 155
   Neutrophil-Derived Cytokines in Rheumatoid Arthritis 156
   Production of Interleukin-8 158
   Production of Other Cytokines 159

8. Cytokine Production by Neutrophils In Vivo 163
Effect of LPS Administration In Vivo
on Neutrophil-Derived Cytokines 163
Other In Vivo Models of Acute Inflammation
Involving Neutrophils 168
Neutrophil-Derived Cytokines During In Vivo Infections 171
Further Experimental Situations of In Vivo
Neutrophil-Derived Cytokines 175

9. Final Remarks and Future Directions 181
Distinct Patterns of Cytokine Release Elicited by Different
Neutrophil Stimuli 181
Differential Abilities of Neutrophils and Mononuclear
Cells to Produce Cytokines 184
Intracellular Control of Cytokine Production 190
Conclusion 191

Index 197

PREFACE

Neutrophils, also known as polymorphonuclear leukocytes (PMN) act as the first line of defense against invading bacteria and other microorganisms. Mature neutrophils have been generally considered to be terminally differentiated cells lacking RNA/protein synthesis capacity because of their short lifespan and of their scarce amounts of organellae for biosynthesis. In recent years, however, it has become increasingly clear that this assumption is incorrect. It has in fact been shown that neutrophils not only synthesize numerous proteins involved in their effector functions but also produce a variety of cytokines. The latter ones are molecules that, among many effects, coordinate the complex network of cellular interactions regulating natural and immune resistance. In view of the broad spectrum of biological activities exerted by cytokines, it can be reasonably speculated that PMN not only play an important role in eliciting and sustaining inflammation, but may also significantly contribute to the regulation of immune reactions.

Unfortunately, although an increasing body of data is already available, there are no available published reviews on this subject. I believe that time is now mature to present our knowledge on this topic in an efficient and organized manner. Therefore, it was my purpose to describe in this book the current knowledge on the production of cytokines by PMN, their molecular regulation and their possible biological and pathophysiological significance.

The book collects and summarizes all the information on the subject, that, to date, has appeared in the scientific literature, and organizes and unifies those data under specific topics. As evidenced from the table of contents, in the first two chapters are summarized the general features of the neutrophil, the general properties of the cytokines, and, cytokine by cytokine, a schematic description of their function, especially in relation to the neutrophil biology. Then (chapter 3), the main features of the neutrophil production of cytokines are described under a personal view. This is followed (chapter 4) by a detailed analysis of the literature regarding the production of each cytokine by polymorphonuclear neutrophils (PMN), one by one. In a new chapter (chapter 5) are described the findings concerning the modulating effects of some immunological factors (IFNγ, IL-4, IL-10, etc.) and other substances, on the production of cytokines by PMN, and the mechanisms of their action, to date known. Subsequently, the regulation of cytokine production in PMN are covered at the level of gene expression (chapter 6). After that, two more chapters follow: the former focuses on the production of cytokines by neutrophils isolated from patients affected by distinct pathologies, the latter dealing with the production of cytokines by neutrophils observed in animal models. The last chapter covers some of the differential aspects between neutrophils and mononuclear leukocytes in their ability to produce specific cytokines, and proposes further considerations and speculations. To make clearer all the concepts dealt with, as many tables summarizing the data as possible, and illustrations, are included. Moreover, the book contains also general information and technical suggestions. Hopefully, this book will serve to update the reader in this area of research, or will be very helpful for those investigators who want to start studying the production of cytokines by neutrophils.

My group has been among the first, and is still, actively involved in studying all molecular aspects of the production of cytokines by human neutrophils. Let me pay a particular tribute to all the researchers from my own lab who have contributed to research in this rapidly-moving field in the past few years: Federica Calzetti, Lucia Meda, Sara Gasperini, Stefano Bonora and Flavia Bazzoni. Numerous colleagues and collaborators have also made valuable contributions, including Filippo Rossi, Miroslav Ceska, Marco Baggiolini, Annalisa D'Andrea, and Giorgio Trinchieri. I want also to thank Maria Cristina Serra for her critical reading of this manuscript. Finally, a particular thanks to P.P. McDonald for his many suggestions and invaluable help in critical editing of most chapters.

Marco Antonio Cassatella
Associate Professor of General Pathology
April 1996

ABBREVIATIONS

ACT D	Actinomycin D
AIDS	Acquired Immunodeficiency Syndrome
Ag	antigen
AM	Alveolar Macrophages
BAL	Bronchoalveolar Lavage
CD30L	CD30 ligand
CGD	Chronic Granulomatous disease
CHX	Cycloheximide
Con A	Concanavalin A
CPPD	Calcium Pyrophosphate Dihydrate
CR1	Complement Receptor Type 1
CR3	Complement Receptor Type 3
CSF	Colony-stimulating factor
DEX	Dexamethasone
EBV	Epstein Barr virus
EC	Endothelial cells
Fcγ	Receptors for the Fc portion of Ig
fMLP	formyl-methionyl-leucyl-phenylalanine
G-CSF	Granulocyte CSF
GM-CSF	Granulocyte Macrophage CSF
GROα	Growth related gene product-α
HIV	Human Immunodeficiency virus

ICE	IL-1 converting enzyme
IFNα	Interferon-α
IFNγ	Interferon-γ
IgG	Immunoglobulin G
IH	immunohistochemistry
IL	Interleukin
IL-1ra	IL-1 receptor antagonist
icIL-1ra	intracellular IL-1 receptor antagonist
IL-1RI	type I IL-1 receptor
IP-10	Interferon Inducible Protein-10
ip	intraperitoneal
ISH	in situ hybridization
it	intratracheal
iv	intravenous
LPS	Lipopolysaccharide
LTβ	Lymphotoxin-β
LTB$_4$	Leukotriene B$_4$
MHC	Major Histocompatibility Complex
M-CSF	Macrophage CSF
MCP-1	Monocyte Chemotactic Protein
MNC	Mononuclear Cells
MPO	myeloperoxidase
MGSA	Melanoma Growth-Stimulatory Activity

MIP	Macrophage Inflammatory Proteins
MSU	monosodium urate monohydrate
NAC	N-acetylcysteine
NK	natural killer
NO	nitric oxide
PAF	Platelet-Activating Factor
PB	peripheral blood
PBMC	Peripheral Blood Mononuclear cells
PCR	polymerase chain reaction
PGE_2	Prostaglandin E_2
PMN	Polymorphonuclear Leukocytes
RA	Rheumatoid Arthritis
RIA	radioimmunoassay
ROI	reactive oxygen intermediates
RSV	Respiratory Syncytial virus
RT-PCR	reverse-transcriptase-PCR
SCF	Stem cell factor
SF	synovial fluid
SNP	sodium nitroprusside
TGF	Transforming growth factor
Th	T helper
TNFα	Tumor Necrosis Factor-α
TNFr-p55	55 kDa TNFα receptor

TNFr-p75   75 kDa TNFα receptor

Y-IgG   *S.cerevisiae* opsonized with IgG

CHAPTER 1

THE NEUTROPHIL

INTRODUCTION

Neutrophils, also known as polymorphonuclear leukocytes (PMN) are the predominant infiltrating cell type present in the cellular phase of the acute inflammatory response. They have been described as mobile arsenals that seek and destroy a variety of targets. Killing mainly occurs via the release of lytic enzymes which are stored in cytoplasmic granules, or through the rapid generation of reactive oxygen intermediates (ROI), such as superoxide anion and hydrogen peroxide. In recent years, however, it has become increasingly clear that PMN are also capable of de novo protein synthesis. It has in fact been shown that neutrophils not only synthesize numerous proteins involved in their effector functions, such as some complement components, Fc receptors, and cationic antimicrobial proteins,[1,2] but also produce a variety of cytokines.[1,3] Since the latter molecules exert a broad spectrum of biological activities, it can be reasonably speculated that, in addition to the inflammatory process, PMN may also significantly contribute to the regulation of many other human responses. Before discussing in detail this topic, I shall describe in this chapter the properties of the neutrophil, its mode of activation and the mechanisms of neutrophil recruitment.

NEUTROPHIL DIFFERENTIATION AND MATURATION

Mature granulocytes are derived from pluripotential stem cells located in the bone marrow (at a rate of 80 million per minute), under the influence of several growth factors named colony-stimulating factors (CSFs), which include multi-CSF (also known as Interleukin-3), Granulocyte and Macrophage CSF (GM-CSF), Granulocyte CSF (G-CSF), and Macrophage CSF (M-CSF). These growth factors exert their differentiation and growth effects on progressively more committed stages of maturation in leukocytes, and can also influence the activities of mature granulocytes, triggered by inflammatory stimuli in

---

Cytokines Produced by Polymorphonuclear Neutrophils: Molecular and Biological Aspects, by Marco A. Cassatella. © 1996 R.G. Landes Company.

vivo.[4] Neutrophils are members of the granulocyte family of leukocytes, which also comprises eosinophils (< 1.5%) and basophils (< 0.5%). All three cell types contain distinct cytoplasmic granules, which are storage pools for intracellular enzymes, cationic protein, receptors and other proteins. The neutrophil granules characteristically do not stain with Wright's stain, hence the name neutrophil. Another morphological peculiarity of the neutrophil is its nucleus: it is polymorphous, and usually consists of three to five sausage-shaped masses of chromatin connected by fine threads.

Neutrophils constitute the most numerous leukocyte type in the blood (normally 50-70% of the total), are 10-20 µM in diameter, and are short lived (a few days) compared to monocytes/macrophages, which may live for months or even years. Within the circulation, PMN exist in two pools which are in a dynamic equilibrium: a circulating pool, and a "marginated" pool; the latter is believed to be sequestered within the microvasculature of many organs. Under pathological conditions such as bacterial infections, the number of circulating neutrophils may increase dramatically (even up to 10-fold), as a result of an accelerated release of neutrophils from the bone marrow, combined with a stimulated maturation of immature neutrophils by CSFs and demargination from the lungs or the spleen. Cell-labeling experiments have shown that the life span of neutrophils in the circulation is short, with a half-life of approximately seven hours. Senescent neutrophils are thought to undergo apoptosis prior to removal by macrophages.[5] This process may also play a role in terminating inflammatory responses, and its importance can be illustrated by the fact that PMN disintegration in vivo would cause the release of their cytotoxic content into the extracellular milieu.

NEUTROPHIL RECEPTORS

Neutrophils have developed a recognition apparatus able to specifically bind a wide range of ligands, suggesting that the interaction of PMN with the external milieu is fundamental for the host response. Since it is beyond the scope of this book to list all the neutrophil receptors so far characterized, I will briefly mention only those which are most likely to be involved in the context of cytokine production. These include: receptors for inflammatory mediators [the anaphylotoxin C5a, Leukotriene B_4 (LTB_4), platelet-activating factor (PAF), substance P], and bacterial formylated peptides; receptors for cytokines [CSFs, IL-1, Tumor Necrosis Factor-α (TNFα), Interferon-γ (IFNγ) etc.]; opsonin receptors, such as those for the Fc portion of immunoglobulins (Fcγ-receptors), namely FcγRI, FcγRII, and FcγRIII, and receptors for the major cleavage fragments of C3 (C3b and C3bi) and C4, called complement receptor type 1 (CR1), and CR3 whose ligand is C3bi; receptors for endothelium, which are a family of at least three different glycoproteins, each consisting of an identical β-subunit (CD18) non-covalently linked to different α-subunits (CD11a, CD11b, CD11c, corresponding, respectively, to LFA-1, MAC-1, and p150,95); receptors for glycoproteins and carbohydrates, such as mannose/fucose glycoproteins, glucan, and Concanavalin A (Con A); receptors for tissue matrix proteins like transferrin, fibronectin, etc.

Following ligation of one or more types of surface receptors on neutrophils, a number of activation steps occur, via the generation of intracellular "second messengers." These steps are biochemical events which mediate the transmission of biological information between membrane receptors and the various effector components involved in such functions as movement, degranulation, metabolic activation, and so forth (Table 1.1 lists some of the potential neutrophil responses to a given ligand). The transduction machinery is mainly, but not exclusively, located in the plasma membrane and is composed of a series of enzymes (adenylate cyclase, kinases, phosphatases, proteases, phospholipases and other enzymes of lipid metabolism) or regulatory proteins (channel proteins, G protein subunits, anchoring proteins), which in turn generate, or whose activities are regulated by, several other messengers (calcium, inositol phosphates, diacyglycerol, phosphatidate, cAMP, etc.). It is not my purpose to dissect the complexity of these systems and their inter-relationships; other references can be consulted for further reading.[6,7]

In view of the various molecules that interact with neutrophils, it is evident that neutrophil receptors are crucially involved in the control of cell activity, for example in particle uptake and ingestion, in chemotactic responses, or in binding to the endothelium of inflamed areas.

Table 1.1. Functional responses of neutrophils to agonists

shape change
adhesion
aggregation
chemotaxis
exocytosis
phagocytosis
respiratory burst
release of arachidonic acid and its derivatives
platelet activating factor synthesis and release
antibody dependent cell cytotoxicity
de novo mRNA and protein synthesis
production of cytokines

NEUTROPHIL MICROBICIDAL MECHANISMS

Two primary processes are utilized by neutrophils to destroy invading pathogens.[8] One involves the release of lytic enzymes and antimicrobial polypeptides contained within intracellular granules, whereas the other process involves the generation of ROI produced by an enzymatic system unique to phagocytic cells, the NADPH oxidase. In addition, PMN appear to produce also reactive nitrogen species.[9] I shall now briefly discuss the biologically active constituents of the two former processes.

Cytoplasmic granules of PMN are heterogeneous, with at least two, and probably several more, different types. One type is the primary (azurophil) granules, thus named because they are the first to appear during hemopoiesis; they constitute the densest granule populations. Primary granules contain lysozyme and myeloperoxidase (MPO), as well as cationic proteins (i.e., bactericidal/permeability-increasing protein/BPI and defensins) which play a direct antimicrobial role at neutral and alkaline pH, proteases and acid hydrolases. MPO is often used as a specific marker for these granules, and is responsible for the characteristic greenish color of neutrophils. The second major type of granules is termed specific or secondary; these granules, which are smaller and less dense than primary granules, stain heavily for glycoproteins. They are known to contain lysozyme, collagenase, vitamin B_{12}-binding protein, lactoferrin, and variety of preformed receptors such as CR3. Finally, there is a heterogeneous variety of granules in addition to those described above, which remain to be fully characterized. Those granules are collectively called tertiary granules, and their primary constituent is gelatinase, a metallo-enzyme.[8]

In addition to granule constitutents, neutrophils are able to produce powerful oxidizing species (ROI) through activation of the NADPH oxidase. This enzyme, dormant in resting cells, is a multiprotein complex consisting of a flavocytochrome-b_{558}, composed of a heavy chain (gp91-phox) and a light chain (p22-phox), and some cytoplasmic components, namely p40-phox, p47-phox, p67-phox, and the GTP-binding regulatory protein, rac,[10,11] which are dissociated in unstimulated PMN. Upon cell stimulation, the cytosolic components and the cytochrome translocate to the plasma membrane and assemble together, thus forming the active enzyme. The primary oxidants formed include the superoxide anion, singlet oxygen, hydrogen peroxide (H_2O_2), as well as metabolic products of H_2O_2. MPO-dependent oxyhalides such as hypochlorous acid (HOCl) are generated by the reaction of H_2O_2 with the abundant Cl^- ions taken up from extracellular fluids; secondary chlorinated amines are generated by the reaction of HOCl with nitrogen-containing compounds.[12] Interestingly, neutrophils contain large reserves of endogenous antioxidants such as glutathione and ascorbate,[13] which protect them from oxidative suicide, for example during phagocytosis.

Remarkably, both granule release and ROI generation can be influenced—either positively or negatively—by a wide variety of mediators, which include cytokines, bioactive lipids and neuroendocrine factors. For instance, the generation of ROI by neutrophils and macrophages in vitro has been shown to be enhanced by previous incubation with IFNγ or lipopolysaccharide (LPS).[14-16] This process, called "priming," allows the phagocytes to acquire a state that enables them to mount a more powerful response once appropriately stimulated. Our studies and those of other investigators (reviewed in ref. 17) have indicated that the modulation by IFNγ of the expression of genes encoding some NADPH oxidase components might be fundamental for the potentiation of neutrophil and macrophage respiratory burst ability. However, the exact molecular mechanisms whereby NADPH oxidase activity is primed have not been elucidated to date. The critical role of NADPH oxidase and its products in host defense is best illustrated by chronic granulomatous disease (CGD). CGD is a rare genetic childhood disease in which phagocytes do not produce ROI, and patients are highly susceptible to bacterial and fungal infections, which can lead to early death.[18] In addition, CGD patients display aberrant inflammatory responses and tissue granuloma formation, the mechanism of which is not known.[19]

In a broader context, the vital role of neutrophils in host defense can be illustrated by the consequences of deficiencies in neutrophil activity. Acquired secondary deficiency of neutrophil numbers (for example, during therapy with cytotoxic drugs) and/or function (for example, following burns or major trauma) is much more common than so-called primary deficiencies (CGD, specific granules deficiency, or adhesion molecules deficiency). The resulting immunodeficiency state is characterized by recurrent infections with bacteria that are usually associated with the formation of pus [including both Gram-positive (*Staphylococci* and *Streptococci*) and Gram-negative species (such as *E. coli*, *Serratia* and *Klebsiella*)]. Patients with neutropenia are also at risk of opportunistic fungal infections with *Candida* and *Aspergillus* species.[8]

THE ROLE OF NEUTROPHILS IN ACUTE INFLAMMATION

When pathogenic agents enter into, or localize in a tissue, neutrophils are recruited to the tissue by a sequence of events beginning with the elaboration of mediators which attract them from the intravascular compartment. Such inflammatory mediators are called chemotactic factors, and include products of complement activation (C5a), lipoxygenase-derived metabolic products (LTB$_4$), certain cytokines called chemokines (IL-8, GROα and others), and bacterial formylated peptides, typified by formyl-methionyl-leucyl-phenylalanine (fMLP). Chemotactic factors derive from numerous sources, including activated plasma components and various cell types such as macrophages and endothelial cells. The local generation of chemotactic mediators eventually promotes the adhesion of the neutrophils to the endothelium, followed by diapedesis (the process of squeezing between endothelial lining cells), migration to the injury site, and activation. Adhesion of leukocytes to the endothelium during inflammation is a multi-step and highly complex phenomenon which requires specific leukocyte-endothelial interactions involving different families of adhesion molecules. The latter include members of the selectin family and their cognate carbohydrate and glycoprotein ligands, which mediate leukocyte deceleration along the vessel wall (a process called "rolling"), as well as the integrins and their cognate immunoglobulin superfamily ligands, which mediate high affinity adhesion of leukocytes to venules.[20] Adhesive receptors expressed by neutrophils may, besides participating in adhesive interaction, transmit signals activating selective leukocyte functions, such as spreading, respiratory burst activity and cytokine gene expression. The importance of these adhesive glycoproteins in neutrophil function in vivo is illustrated by the fact that individuals who genetically lack the leukocyte adhesive proteins CD11b and CD11c, display an abnormally high susceptibility to bacterial infections.[20]

After the process of emigration, in which neutrophils undergo shape changes and mobilize their secondary granules to the cell surface, cells arrive to infection foci, where they adhere to extracellular matrix components such as laminin and fibronectin, and begin to react with the etiopathogenetic agent. Although granulocytes do not apparently show any particular specificity for antigens, they nevertheless play an important role in host protection against microorganisms. The elimination of the latter takes place through phagocytosis, which is triggered upon the binding of opsonized microorganisms through receptors for complement fragments or for antibodies, or through nonspecific glycosylated receptors that recognize certain lectins on target microorganisms. The foreign particle is then internalized within cytoplasmic phagosomes which eventually fuse with granules, thereby forming phagolysosomes. This in turn results in the subsequent release of proteolytic enzymes and other bactericidal components into the phagolysosome. A substantial amount of these factors may also be directly released into the external milieu (exocytosis), thus causing damage to the connective tissue and nearby cells. At the same time, a dramatic increase of oxygen consumption occurs, through activation of the hexose monophosphate shunt and generation of ROI, that are released into the phagolysosome or, alternatively, outside the cell. The intravacuolar pH is first subjected to a rapid alkalinization, due to both protonation of superoxide and consumption of hydrogen ions in the dismutation reaction, followed by a slow and progressive acidification, due to the activation of an Na^+/H^+ antiporter. While all of the above-mentioned events are necessary and essential for neutrophil-mediated microbicidal activity and digestion of engulfed particles, the same mechanisms can also be involved in other effects of PMN, such as tumoricidal activity, cytotoxicity, tissue matrix injury, amplification of the inflammatory process, and priming of the tissue-healing processes.

Once initiated, the acute inflammatory response may be rapidly amplified by the neutrophils themselves, as well as by locally produced mediators. Neutrophils are for example potent producers of arachidonic acid-derived bioactive lipids, such as LTB_4. Other mediators are cytokines (i.e., G-CSF, GM-CSF, IL-1, TNFα and IFNγ), which can influence for example neutrophil survival,[5] and prime them for their effector functions.[17] Cytokines may also initiate a systemic response, with synthesis of acute-phase proteins by the liver, accelerated myeloid proliferation in the marrow, neuro-endocrine modifications, and so forth. All these events lead to a massive leukocytosis, and through the "priming" of circulating neutrophils, to a marked potentiation of their cellular functions.

NEUTROPHILS IN HUMAN DISEASES

While the activities of neutrophils are normally beneficial and protective as a whole, they can also cause extensive tissue necrosis under certain conditions, especially if PMN are inappropriately stimulated.

There even exist certain situations in which neutrophils themselves may play a major pathogenic role, through a deregulated control of their effector functions, or through their continuous activation. Examples include many chronic inflammatory conditions, glomerulonephritis, inherited deficiency of α1-antitrypsin (the major physiological inhibitor of neutrophil elastase), in which a severe pulmonary emphysema with alveolar destruction is developed, ischemia-reperfusion injury,[21] the adult respiratory distress syndrome,[22] and immune-complex diseases such as type III hypersensitivity, in which PMN are activated through their Fcγ-receptors by immune complexes. Based on many studies, neutrophil oxidants and proteases, acting individually as well as in concert, appear to be responsible for much of the tissue injury occurring under the above conditions. Future research should make it possible to develop selective strategies for inhibition of neutrophil-mediated tissue injury without interfering with the cells' ability to kill invading microorganisms.

REFERENCES

1. Loyd AR, Oppenheim JJ. Poly's lament: the neglected role of the polymorphonuclear neutrophil in the afferent limb of the immune response. Immunol Today 1992; 13:169-172.
2. Waksman Y, Golde DW, Savion N et al. GM-CSF enhances cationic antimicrobial protein synthesis by human neutrophils. J Immunol 1990; 144:3437-3443.
3. Cassatella MA. The production of cytokines by polymorphonuclear leukocytes. Immunol Today 1995; 16:21-26.

4. Vose JM, Armitage JO. Clinical applications of hematopoietic growth factors. J Clin Oncol 1995; 13:1023-1035.
5. Savill JS, Fadok V, Henson P, et al. Phagocyte recognition of cells undergoing apoptosis. Immunol Today 1993; 14:131-136.
6. Alberts B, Bray D, Lewis J, Raff M, Roberts K, Watson JD. Molecular biology of the cell. 3rd ed. New York, London: Garland Publishing, 1995:721-770.
7. Lodish HL, Baltimore D, Berk A, Zipursky SL, Matsudaira P, Darnell J. Molecular cell biology. 3rd ed. New York: W.H. Freeman and Company, 1995:850-922.
8. Segal AW, Walport MJ. Neutrophil leukocytes. In: McGee J, Isaacson PG, Wright NA, eds. Textbook of Pathology. Oxford University Press, 1994:321-329.
9. McCall TB, Boughton-Smith NK, Palmer RMJ et al. Synthesis of nitric oxide from L-arginine by neutrophils: release and interaction with superoxide. Biochem J 1989; 261:293-296.
10. Chanock SJ, El Benna J, Smith RM et al. The respiratory burst oxidase. J Biol Chem 1994; 269:24519-24522.
11. McPhail L. SH3-dependent assembly of the phagocyte NADPH oxidase. J Exp Med 1994; 180:2011-2015.
12. Weiss SJ. Tissue destruction by neutrophils. N Engl J Med 1989; 320:365-376.
13. Eaton JW. Defenses against hypoclorous acid: parrying the neutrophil's rapier thrust. J Clin Lab Med 1993; 121:197-198.
14. Berton G, Zeni L, Cassatella MA et al. Gamma interferon is able to enhance the oxidative metabolism of human neutrophils. Biochem Byophis Res Commun 1986; 138:1276-1282.
15. Nathan CF, Murray HW, Wiebe ME et al. Identification of IFNγ as the lymphokine that activates human macrophage oxidative metabolism and antimicrobial activity. J Exp Med 1983; 158:670-689.
16. Guthrie LA, McPhail LC, Henson PM et al. Priming of neutrophils for enhanced release of oxygen metabolites by bacterial lipopolysaccharide. J Exp Med 1984; 160:1656-1671.
17. Cassatella MA, Bazzoni F, Aste Amezaga M et al. Studies on the gene expression of several NADPH oxidase components. Biochem Trans 1991; 19:63-67.
18. Gallin JI, Malech, HL. Update on chronic granulomatous diseases of childhood. J Am Med Assoc 1990; 263:1533-1537.
19. Ament M, Ochs H. Gastrointestinal manifestations of chronic granulomatous disease. N Engl J Med 1973; 288:382-387.
20. Springer TA. Traffic signals on endothelium for lymphocytes recirculation and leukocyte emigration. Annu Rev Physiol 1995; 57:827-872.
21. Ricevuti G, Mazzone A, Pasotti D et al. Role of granulocytes in endothelial injury in coronary heart disease in humans. Atherosclerosis 1991; 91:1-14.
22. Martin TR, Pistorese BP, Hudson LD et al. The function of lung and blood neutrophils in patients with adult respiratory distress syndrome. Am Rev Resp Dis 1991; 144:254-262.

CHAPTER 2

THE CYTOKINES

A DEFINITION OF CYTOKINE

Cytokine is a term that encompasses a wide variety of soluble polypetides and glycoproteins that are in many ways similar to hormones (but that are not necessarily endocrine-derived), and that can affect cell function. Most are secreted, but some can be expressed on the cell membrane, while others are held in reservoirs in the extracellular matrix. Cytokines act primarily in the local milieu, usually functioning in an autocrine and paracrine fashion, although many have been found at low static levels in the circulation. Generally speaking, cytokine receptors have a very high affinity for their ligands; consequently, cytokines are extremely potent at low concentrations. Their potency is partly due to cascades and networks, whereby one cytokine leads to the secretion of others, often resulting in amplification loops. Conversely, there may also be reciprocal cytokine interactions resulting in negative feedback loops.

Cytokines participate in a wide variety of biological responses and are increasingly viewed as pivotal elements in enabling communication among different cell types. When first discovered, cytokines were presumed to have single cell sources, single targets, and single specific functions. Accordingly, products of monocytes/macrophages were called monokines, lymphokines came from lymphocytes, and growth factors affected cell growth. The term interleukin (IL) was introduced to foster the view that leukocytes regulated leukocyte functions through these secreted molecules. The list of cytokines soon expanded beyond the interleukins to include the interferons, tumor necrosis factors, and various colony-stimulating factors (CSFs), and chemotactic cytokines ("chemokines") (Table 2.1). It was also rapidly realized that cytokines can have multiple cellular sources, multiple targets, a broad (and sometimes) overlapping range of biological activities, and that they function in cross-talk networks as a complex signaling language. This functional networking, in combination with the pleiotropic actions of cytokines,

---

Cytokines Produced by Polymorphonuclear Neutrophils: Molecular and Biological Aspects, by Marco A. Cassatella. © 1996 R.G. Landes Company.

Table 2.1. Cytokines

Interleukins	Tumor Necrosis Factors (TNF)
Interleukin-1α	Cachectin (TNFα)
Interleukin-1β	Lymphotoxins (TNFβ$_1$, TNFβ$_2$)
Interleukin-1 receptor antagonist	
Interleukin-2	
Interleukin-3	**Colony Stimulating Factors (CSF)**
Interleukin-4	Granulocyte CSF
Interleukin-5	Macrophage CSF
Interleukin-6	Granulocyte Macrophage-CSF
Interleukin-7	
Interleukin-9	
Interleukin-10	**Growth Factors**
Interleukin-11	Keratinocyte Growth Factor
Interleukin-12	Epidermal Growth Factor
Interleukin-13	Fibroblast Growth Factors
Interleukin-14	Hepatocyte Growth Factor
Interleukin-15	Insulin-like Growth Factors
Interleukin-16	Platelet-Derived Growth Factors
Interleukin-17	Vascular Endothelial Growth Factor
	Others
Chemokines	**Transforming Growth Factors (TGF)**
C-X-C Branch (IL-8, etc.)	TGFα
C-C Branch (MCP-1, etc.)	TGF-β1, -β2, -β3, -β4, -β5
C Branch (Lymphotactin)	Activins
Interferons	**Neurotrophic Factors**
Interferon-α	Nerve Growth Factor (NGF)
Interferon-β	Brain Derived Neurotrophic Factor (BDNF)
Interferon-γ	Ciliary Neurotrophic Factor (CNTF)
	Neurotrophins (NT)
Various	
Erythropoietin	
Leukemia Inhibitory Factor	
Oncostatin M	
Stem Cell Factor/Kit Ligand	
Angiogenin	

naturally leads to a certain redundancy in the biological effects of the various cytokines. In this respect, gene deletion experiments have established that few individual cytokines are absolutely essential for life, presumably because the remaining cytokines can compensate for the loss of a given one.

In general, cytokines function as growth and differentiation factors, alarm signals for the inflammatory response, chemotactic factors for leukocytes, modulators of immune cell function, and tissue remodelers. Cytokine biology is an essential part of how the body mounts physiological responses, and how these responses are controlled. Most cytokines are not constitutively expressed in adult animals. On the contrary, they are rapidly produced in response to stimulation by infectious agents (or their derived products such as endotoxin), inflammatory mediators, mechanical injury, and cytokines themselves.

Cytokines bind to specific receptors on the surface of target cells and then generate intracellular signal transduction and second messenger pathways. The receptors for many cytokines have now been cloned and analysis of their primary structures has enabled their regrouping into superfamilies, based on common homology regions. The main cytokine receptor superfamilies are the hematopoietic receptor superfamily or cytokine receptor superfamily type I [recognizing IL-2, IL-3, IL-4, IL-5, IL-6, IL-7, IL-9, IL-13, Granulocyte Colony-Stimulating Factor (G-CSF), Granulocyte Macrophage (GM-CSF), Ciliary Neurotrophic Factor (CNTF), Oncostatin (OSM), LIF, and erythropoietin], the immunoglobulin receptor superfamily, the protein tyrosine kinase receptor superfamily (recognizing M-CSF, EGF, PDGF, FGF and TGFα), the nerve growth factor/tumor necrosis factor receptor superfamily (recognizing NGF, TNFα, Lymphotoxin-β), the interferon receptor superfamily, also known as the cytokine receptor superfamily type II (recognizing IFN-α/-β/-γ, and IL-10), and the G-protein-coupled seven transmembrane spanning receptor superfamily (recognizing the chemokines).

Specific details about the various cytokines are outlined below, with special emphasis on their actions towards neutrophils.

INTERFERON-α (IFNα)

The α interferons (IFNα) are a family of closely related inducible secreted proteins of 165-172 amino acids, which are important not only in defense against a wide range of viruses, but also in the regulation of the immune responses and in hematopoietic cell development.[1] Whilst there is only one IFNγ and one IFNβ gene, there are at least 24 different genetic loci for the α interferons of which 15 correspond to functional genes. Unusually for cytokines, both IFNα and IFNβ genes lack introns, and due to their extensive homologies in their amino acid sequences they have been grouped as type I interferons.[1,2] To date, two IFNα receptors have been cloned, the distributions of which are ubiquitous.[3,4] The two IFNα receptor proteins are members of the class II cytokine receptor family which also includes the tissue factor receptor. It is not known if these two receptors are expressed independently on the cell surface or are associated with each other in an IFNα receptor complex.

IFNα is produced by most cells (but mainly by leukocytes including monocytes/macrophages, B cell lines, and neutrophils) in early response to viral infections, following stimulation with natural or synthetic double-stranded RNA such as poly (I:C), as well as in response to bacterial, mycoplasma and protozoan infection, and to certain cytokines.[2] In addition to inducing a potent antiviral state in uninfected cells, IFNα, at higher concentrations, has antiproliferative activity against both normal and tumor cells.[1,2] They are important in regulating cell mediated as well as some aspects of humoral immunity, since they can enhance, for example, the expression of class I MHC gene products,[5] natural killer cell activity,[6] and macrophage activation.[7] In addition, it has also been reported that IFNα augments the antifungal activity of *Candida*-phagocytizing neutrophils.[8] Several gene products are induced by IFNα, including 2-5A synthetase, p68 kinase, the Mx family of proteins, etc.[1] The activities of IFNα have been exploited therapeutically for the treatment of viral infections and tumors, such as, for example, chronic hepatitis B and C, and hairy cell leukemia, respectively.[9] Defects in IFNα production have been associated with diminished host resistance in both human and murine systems.[10]

INTERFERON-γ (IFNγ)

IFNγ is a pleiotropic cytokine involved in the regulation of nearly all phases of immune and inflammatory responses, by acting on the growth, differentiation and activation of T cells, B cells, macrophages, natural killer (NK) cells and other cell types such as neutrophils and fibroblasts.[11,12] IFNγ, also known as type II interferon or immune interferon, is a glycoprotein which is secreted from T cells and NK cells.[11,12] It has weak antiviral and antiproliferative activity, and in some conditions potentiates the antiviral and anti-tumor effects of IFNα/β.[11,12] The IFNγ receptor is a complex of a high affinity IFNγ-binding chain and a second accessory protein required for signal transduction, both members of the class II cytokine receptor family.[13] The IFNγ receptor is expressed on a wide variety of hematopoietic cells, excluding erythrocytes, but also on many somatic and tumor cells.[13] IFNγ induces the expression of many key molecules, including MHC class I and II antigens, cell surface ICAM-1, nitric oxide synthase and some cytokines.[11,12] IFNγ is responsible for inducing nonspecific cell mediated mechanisms of host defense, through its ability to activate macrophages to become tumoricidal and to kill intracellular parasites.[11,12] Other than being a macrophage activating factor, evidence has been accumulated that IFNγ is also a neutrophil-activating factor.[14] IFNγ induces the surface expression of FcγRI,[15] enhances neutrophil antibody dependent cell cytotoxicity (ADCC),[16] potentiates the generation of ROI and granule release in response to different stimuli,[17,18] and modulates neutrophil microbicidal activity.[14] As described in chapter 5, IFNγ modulates also neutrophil production of cytokines. Clinically, IFNγ has been shown to alleviate the symptoms of Chronic Granulomatous disease[19] and to exhibit potential benefits in the treatment of infectious disease and neoplasia.[11,12]

TRANSFORMING GROWTH FACTOR-β (TGFβ)

Transforming growth factor β (TGFβ) was originally discovered as a secreted factor that induced malignant transformation in vitro.[20] TGFβ comprises a highly homologous family of structurally related 25 kDa disulfide-linked dimeric proteins that are secreted by nearly all cell types, and that modulate a wide range of activities.[20] Five highly conserved isoforms of TGFβ encoded by separate genes and sharing considerable structural and sequence homology have been described, three of which, TGFβ1, TGFβ2, and TGFβ3, are expressed in mammals.[21] The mature processed proteins share 70-80% amino acid sequence identity, bind to the same receptors, induce similar responses, and for most of their activities are interchangeable, even though TGFβ1 knockout mice experiments indicated a limited compensation for other TGFβ isoforms, thus arguing simple functional redundancies.[22,23] TGFβ is usually secreted in a biologically inactive form (except in the case of neutrophils), unable to bind to the TGFβ receptor and to exert its diverse array of biological activities.[24] Generation of active TGFβ is obtained in vitro via enzymatic (plasmin, cathepsin, glycosidase), thermal or acid treatment.[21,20] It is conceivable, but not proved, that the regulation of the biological effects of TGFβ is also attained by the activation of the latent form in the micromilieu of the target cells rather than by enhanced synthesis and/or secretion. TGFβ is a multifunctional cytokine capable of a variety of immunologic effects, many of which are in opposition.[20,21] The effects of TGFβ are dictated by the target cell type, its state of differentiation and the presence of regulatory molecules. Not only does it regulate growth, as evident by its name, but also mediates far-ranging biological processes including inflammation and host defense, development, tissue repair and tumorigenesis. For example, TGFβ orchestrates leukocyte recruitment and activation.[25,26] TGFβ has been demonstrated to modulate the immune response by affecting proliferation, activation state, and differentiation of immune cells that participate in innate, nonantigen (Ag)-specific, as well as Ag-specific, immunity. Stimulatory as well as inhibitory effects have been reported. In vitro, TGFβ has been shown to oppose the proinflammatory effects of both TNFα and IFNγ,[27] to inhibit activation of macrophages,[28] to modulate production and effects of monocyte proinflammatory cytokines,[29] to modulate expression of surface immunoregulatory molecules such as HLA-DR determinants[30] and receptors for the Fc fragment of IgG (FcγRIII/CD16).[31] TGFβ also suppresses IFNγ-stimulated nitric oxide (NO) production by murine peritoneal macrophages.[32] TGFβ facilitates healing by promoting fibroblast recruitment and matrix synthesis.[33] As a general, but by no means exclusive rule, TGFβ serves as a conversion factor, converting an active inflammatory site into one dominated by resolution and repair. Among TGFβ's multiple actions is strong induction of extracellular matrix deposition by simultaneously stimulating the production of matrix proteins, inhibiting proteases that degrade matrix, and modulating the expression of matrix receptors on the cell surface.[20,21,27] TGFβ's potency in stimulating matrix deposition has offered hope that it might be used clinically to enhance healing of problem wounds.[34] At the same time overproduction of TGFβ has now been linked to the pathogenesis of numerous experimental and human fibrotic disorders.[35] In fact, upsetting the delicate balance of TGFβ that dictates these events may have pathologic consequences. Thus TGFβ is a double-edged sword with both therapeutic and pathological potential.

Receptors for TGFβ were shown to be expressed on the surface of neutrophils.[26] Femtomolar concentrations of TGFβ1 have been shown to induce neutrophil chemotaxis,[26,36] but higher TGFβ concentrations suppressed this chemotactic response.[36] More recent in vitro studies showed that all TGFβ isoforms possess very strong chemotactic activities on polymorphonuclear leukocytes (PMN) themselves (being TGFβ2 > TGFβ1 > TGFβ3).[37] Contrary to other chemotactic factors such as formyl-methionyl-leucyl-phenylalanine, fMLP, or IL-8, TGFβ does not elicit degranulation and reactive oxygen intermediates (ROI) release from neutrophils.[26,37] Moreover, TGFβ does not seem to influence specific functional responses of neutrophils to activatory factors such as fMLP or IL-6.[36,38]

TUMOR NECROSIS FACTOR-α (TNFα)

TNFα is a paracrine and endocrine mediator with potent immunomodulatory and proinflammatory properties, which plays a major role in host defense.[39] TNFα is a 17 kDa protein, and can be also expressed as a membrane protein attached by a signal anchor transmembrane domain in the propeptide, that can be processed by a matrix metalloproteinase.[40] TNFα belongs to a large family of membrane-anchored and soluble cytokines including Lymphotoxin-β (LTβ), FAS ligand, CD40 ligand (CD40L), CD30 ligand (CD30L), CD27 ligand, and 4-1BB ligand, all of them with their receptors, being molecules centrally involved in T cell immunity.[41] TNFα is produced mainly by activated monocytes and macrophages in response to many stimuli, but several other cell types including B and T cells, mast cells, fibroblasts and neutrophils can produce small amounts also.[39] TNFα exerts its cellular effects by binding to two types of TNFα receptors, the TNFr-p55 and the TNFr-p75.[42] Both receptors can be proteolytically cleaved from the cell surface and thus form soluble proteins that retain the capacity to bind TNFα with high affinity. Both receptors are members of the NGF receptors/TNF receptors superfamily, which include CD27, CD30, CD4, FAS antigen, and others, and are present on nearly all cell types.[42] The original interest in TNFα was based on its antitumoral activity, since it was first identified as a factor found in the serum of *Bacillus Calmette-Guerin*-treated mice which caused hemorragic regression of transplanted tumors in the same mice.[43] TNFα is in fact selectively cytotoxic for many transformed cells, especially in combination with IFNγ. In fact, many of the actions of TNFα occur in combination with other cytokines as part of the cytokine network. TNFα also exerts a variety of strong proinflammatory effects on different cell types, including induction of catabolic states in adipocytes, induction of endothelial cell adhesion molecules, induction of plasminogen activator and its inhibitor, induction of other cytokines, and so forth.[39,44] As a result, TNFα contributes significantly to vasodilatation, thrombosis, leukocyte recruitment, bone resorption, matrix degradation in cartilage, changes in liver metabolism, pannus formation and cachexia, consistent with an important role in many disease states.[44,45] In addition, TNFα has been shown to be a primary mediator of the pathology seen in endotoxic shock,[45] as animals with a TNFr-p55 receptor gene knockout are resistant to septic shock.[46] Excessive or prolonged production of TNFα is a feature of several important diseases, e.g., septic shock, rheumatoid arthritis, Crohn's disease, and multiple sclerosis.[44,45] Recent clinical trials have indicated that blocking the action of TNFα with specific monoclonal antibodies can have significant benefit in rheumatoid arthritis[47] and Crohn's disease.[48] The cellular targets of TNFα include monocytes, macrophages, lymphocytes, eosinophils and neutrophils.[44,45] TNFα is a potent activator of neutrophils, as it increases adherence of neutrophils to vascular endothelium,[49,50] enhances neutrophil phagocytosis,[51,52] and antibody-dependent cell cytotoxicity,[53] triggers neutrophil degranulation[54] and release of reactive oxygen metabolites.[52,55] Moreover, under certain conditions, TNFα is also a very potent priming agent for neutrophil functions.[51]

THE INTERLEUKIN-1 SYSTEM (IL-1α/β AND IL-1RA)

IL-1 is the term for two polypeptide mediators (IL-1α and IL-1β) which are among the most potent and multifunctional cell activators described in immunology and cell biology. The spectrum of action of IL-1 encompasses cells of hematopoietic origin, vessel wall elements, cells of mesenchymal, nervous and epithelial origin.[56] IL-1 is an important mediator of the host defense response to injury and infection, and has both protective and proinflammatory effects. The production and action of IL-1 are so important that they are regulated by multiple control pathways, some of which are unique to this cytokine, in a so-called "IL-1 system." This IL-1 system consists of the two agonists, IL-1α and IL-1β, a specific activation enzyme (IL-1 converting enzyme, ICE), a receptor antagonist (IL-1ra) produced in different isoforms,[57] and two surface binding molecules, the type I and type II IL-1 receptors (IL-1RI, IL-1RII).[58] A wide variety of cells, including monocytes, macrophages, dendritic cells, lymphocytes, NK cells, endothelium, fibroblast, glial cells, keratinocytes, and neutrophils are able to secrete IL-1.[56] IL-1α and IL-1β are both synthesized as precursor molecules of 30 to 35 kDa.[56] In human monocytes and murine macrophages, IL-1β is cleaved to its biologically active 17 kDa form by the action of an enzyme, ICE, which is a cysteine protease.[59] Neither

IL-1α nor IL-1β contain the characteristic N-terminal signal sequence that target secreted proteins to the endoplasmic reticulum and export. Thus atypical pathways must be responsible for its export from the cell. A vast phenomenological literature on the activities of IL-1 exists, and it is almost impossible to list them. IL-1 affects the hematopoietic system at various levels, from immature precursors to mature myelomonocytic and lymphoid elements. In particular, IL-1 costimulates T cell proliferation in the classic costimulator assay,[56] induces cytokine production in monocytes,[56] prolongs the in vitro survival of PMN by blocking apoptosis,[60] activates endothelial cells in a proinflammatory, prothrombotic sense,[61] and stimulates the release of corticotropin-releasing hormone (CRH) by the hypothalamus, ultimately causing a release of corticosteroids in the bloodstream by the adrenals.[62] Furthermore, IL-1 is a key mediator of the acute phase response,[63] is the main endogenous pyrogen, an activity shared with other cytokines including TNFα and IL-6,[63] and so forth. In vivo, IL-1 induces hypotension, fever, weight loss, neutrophilia and acute phase response.[56] IL-1 has a number of local effects that have been termed catabolic, and plays a role in destructive joint and bone diseases,[56] other than having important toxic action for insulin-producing β cells in Langerhans islets.[64] In addition, many observations support a role of IL-1 in the pathogenesis of many diseases.[56]

Recent evidence suggests that the biologic effects of IL-1 are regulated by natural inhibitors, including IL-1ra.[57] The IL-1 receptor antagonist (IL-1ra) has been characterized as a 23 to 25 kDa glycosylated protein isolated either from supernatants of blood monocytes on adherent IgG[65] or from the urine of febrile patients[66] and children with juvenile rheumatoid arthritis (RA).[67] In general, IL-1ra is made by the same cells that secrete IL-1, and is an important physiological regulator.[57] IL-1ra is an anti-cytokine that specifically inhibits the proinflammatory actions of IL-1 by binding to IL-1RI and IL-1RII located on a target cell, without initiating any signal transduction.[68] IL-1ra exhibits no IL-1 bioactivity,[56] but antagonizes IL-1 effects.[69] Different structural variants of IL-1ra have been described, the secreted form(s) produced by mononuclear cells, and two intracellular isoforms (icIL-1ra) found in keratinocytes and other epithelial cells,[70] as well as in myelomonocytic lineage.[71,72] The biological function played by the cell-associated fraction of IL-1ra remains largely obscure.

The effects of IL-1α and IL-1β on neutrophil functions are still controversial,[73-76] and differences between the effects of both types of IL-1 have not been elucidated yet. A very recent paper however,[77] reported that both IL-1α and IL-1β alone triggered superoxide anion release in a dose-dependent manner, and together primed human neutrophils for enhanced release of superoxide anion stimulated by chemotactic peptide, chemokine and plant lectin. All these effects were completely inhibited by IL-1ra.[77]

INTERLEUKIN-2 (IL-2)

IL-2 is a 15.5 kDa glycoprotein secreted primarily by activated T lymphocytes, and is well recognized as a growth and differentiation factor for the same T lymphocytes.[78] The de novo synthesis and secretion of IL-2 is triggered primarily by antigen- or mitogen-induced activation of mature T lymphocytes, which in turn, promotes the clonal expansion of antigen-specific effector T cells, in its various forms. IL-2 also delivers signals for both differentiation and proliferation for a variety of hematopoietic cell types, for example playing a role in thymus development.[78] Natural killer (NK) cells and peripheral blood monocytes display receptors for IL-2, and their activities are also widely influenced by this cytokine.[78]

INTERLEUKIN-4 (IL-4) AND IL-13

IL-4 is an 18-20 kDa glycoproteic pleiotropic cytokine derived from T cells and mast cells with multiple biological effects on B cells, T cells and many nonlymphoid cells including monocytes, endothelial cells and fibroblasts.[79] IL-4 binds to a receptor that is a complex of molecules consisting of at least of two chains: a high affinity IL-4 binding chain (α chain), and the IL-2R γ-chain, also known as the common γ-chain (γc).[80] The many immunoregulatory properties of IL-4 include secretion of IgG1, IgG4 and IgE by human B cells due to a IL-4-induced isotype switching,[81] and induction of the subpopulation of helper T cells (designated Th2 cells) which regulate humoral immunity, eosinophilia, and inflammatory macrophage deactivation.[82] Both properties have been confirmed in vivo in animal models. Mutant mice in which the IL-4 gene has been disrupted by homologous recombination are completely unable to produce IgE, and have impaired IgG1 development.[83] In some infectious disease models, administration of anti-IL-4 at the time of infection will divert the ensuing response away from Th2 cells toward the Th1 subpopulation of T helper effectors.[84] Other in vitro properties comprehend growth co-stimulation for B cells, T cells, mast cells, myeloid and erythroid progenitors, and inhibition of inflammatory mediator release from activated monocytes/macrophages.[85] IL-4 also exhibits striking anti-tumor activity in a variety of animal models;[86] this seems to be due to the rapid tumor infiltration by cytotoxic eosinophils.[86] While in vitro IL-4 has been demonstrated to have a dual role of either enhancement or suppression of a number of monocyte functions, very few reports which address its action on neutrophils. For instance, while it enhances monocyte/macrophage expression of MHC molecules,[87] and tumoricidal activity,[88] it suppresses the generation of ROI and of a number of cytokines.[89,90] In neutrophils, IL-4 has been found to induce PMN lysozyme release and migration, to elevate phagocytic activity, and to enhance the respiratory burst.[91,92]

IL-13 is a T-cell derived cytokine, which shares with IL-4 approximately a 25-30% amino acid homology and many biologic activities including the induction of IgE production by B cells.[93] This might be due to the fact that although the receptors for IL-4 and IL-13 are distinct, one subunit is in common.[94,95] IL-13 prolongs the survival of monocytes and inhibits their production of inflammatory cytokines, induces the expression of CD23 in, and promotes proliferation of, B cells.[93]

INTERLEUKIN-6 (IL-6)

IL-6 is a multifunctional cytokine that exerts multiple effects on many different types of target cells.[96,97] It was originally identified as a T cell derived lymphokine that induces the final maturation step of B lymphocytes into antibody-producing cells.[97,98] Other common alternative names of this cytokine were Interferon-β2 (IFNβ2) and hepatocyte stimulating factor (HSF).[99] IL-6 plays an important role in the regulation of the immune response by modulating the functions of T and B lymphocytes and priming endothelial cells.[96-98] It is a terminal differentiation factor for B cell maturation, a growth factor of myeloma, plasmacytoma and hybridoma cells, T cells, mesangial cells and megakaryocytes, and also a pro- and anti-inflammatory factor.[96-98] Sharing several activities with IL-1 and TNFα, IL-6 is a potent inducer of the acute-phase response in the liver, of fever, and can stimulate also the proliferation of early hematopoietic progenitor cells, from which osteoclastic cells take origin.[96-98] The IL-6 receptor is found on both lymphoid and nonlymphoid cells and consists of at least two chains, both of which belong to the hematopoietic cytokine receptor family: an α-chain of molecular weight 80 kDa, the ligand-binding receptor (IL-6R), and a β chain of 130 kDa that is involved in the signaling function.[96] The β chain does not bind IL-6 unless the α chain is present; binding of IL-6 to IL-6R triggers the association of IL-6R and gp130, forming a high affinity IL-6 binding site, and gp130, in turn, transduces the signal.[96] IL-6 is synthesized by an enormous variety of cells, including fibroblasts, T and B cells, macrophages, endothelial cells, astrocytes and keratinocytes,[96-98] and when purified from natural sources it behaves as a single chain of 21-28 kDa.[99] Deregulated production of this cytokine has been implicated in the pathogenesis of several diseases, including rheumatoid arthritis, multiple myeloma, Castleman's disease,[96-98] and postmenopausal osteoporosis.[100,101] Moreover, IL-6 overproduction was shown to trigger plasmacytoma formation in transgenic mice.[102] IL-6 is one of the cytokine that is released early in the course of systemic infections. In experimental lethal bacteriemia in baboons, the appearance of IL-6 in the circulation shortly follows that of TNFα.[103] In addition, elevated serum concentrations of IL-6 are found in the majority of patients with sepsis and show a strong positive correlation with mortality rates,[104] but it is still uncertain to what extent IL-6 is involved in tissue injury in sepsis. Interestingly, in IL-6-deficient mice, which are viable and fertile,[100] several aspects of the acute host response to turpentine, including the liver acute phase reaction, anorexia, loss of body weight, and hypoglycemia, are severely reduced, but the same response after treatment with LPS is only slightly modified, demonstrating that in the latter case IL-6 is dispensable.[105,106]

The effects of IL-6 on neutrophils are controversial.[107] The fact that neutrophils express the IL-6 receptor suggests a regulatory role of IL-6 on neutrophil functions.[108] In addition, further evidence for a potential role on neutrophil functions are the reported priming effects of IL-6 on the respiratory burst,[109] and the IL-6-stimulatory effects associated with the release of lactoferrin, β-glucuronidase, lysozyme,[110] and elastase.[38]

INTERLEUKIN-8 (IL-8)

Migration to and local accumulation of leukocytes at the sites of inflammation are important steps in the reaction cascade during infectious or inflammatory responses. A key mediator for the migration of neutrophils from the circulation is IL-8. This inflammatory cytokine belongs to a recently discovered group of cytokines which are related in terms of both sequence and genomic structures. These have been called by different workers the "small cytokine" family, intercrines or chemokines, the last name now being their agreed title.[111] A partial list of chemokines is shown in Figure 2.1. All these proteins are around 8-10 kDa in size, are basic heparin-binding proteins, act on various inflammatory cell types, and may be separated into subgroups based on whether the first two of the four position-invariant cysteine residues, common among the various primary sequences, are separated by another amino acid (C-X-C subfamily, α-chemokines), or in adjacent positions (C-C subfamily, β-chemokines). Recently, it has been identified a T cell chemoattractant, named lymphotactin, with sequence homology with members of both the α- and β-chemokine subfamilies, but having only two of the four conserved cysteine residues.[112,113] Lymphotactin may therefore be representative of a third subfamily of chemokines (Fig. 2.1). Amino acid analysis has shown that these proteins contain leader sequences suggesting that they are secreted by conventional pathways.[114,115]

IL-8 has been the most extensively studied at both the molecular and biochemical level, and is the prototyte of the α-chemokines, which include among the other members, the IL-8 related proteins GROα/MSGA, GROβ and GROγ (see below), the neutrophil activating peptide 2 (NAP-2), and Interferon Inducible Protein-10 (IP-10).[114,115] All of these proteins are highly stable, not easily denatured by heat, extreme pH changes, or proteases.[115] Apart from IP-10, all α-chemokines act mainly on neutrophils to trigger chemotaxis, respiratory burst, degranulation, and adhesion to endothelial cells.[114,115] IL-8 was originally isolated from stimulated monocytes,[116,117] and was called by different names, namely neutrophil attractant/activating protein (NAP-1), mono-

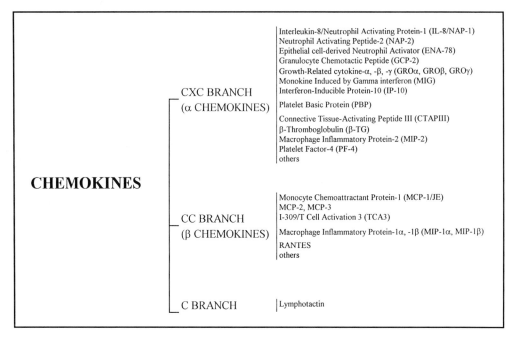

Fig. 2.1. Chemokines.

cyte derived neutrophil activating peptide (MONAP), monocyte derived neutrophil chemotactic factor (MDNCF), neutrophil activating factor (NAF), and granulocyte chemotactic protein (GCP). IL-8 is expressed in response to inflammatory stimuli and is secreted by a variety of cell types, including lymphocytes, epithelial cells, keratinocytes, fibroblasts, endothelial cells (EC), smooth muscle cells, and granulocytes.[114,115] Although the IL-8 gene encodes a 99 amino acid precursor molecule which contains a 20 amino acid signal sequence, the m.w. of secreted cytokine is different, depending on the cell source. EC or fibroblasts predominantly release the 77 amino acid isoform, whereas monocytes mainly produce the 72 amino acid IL-8 isoform.[118] Two IL-8 receptors termed A and B have been cloned, one of high affinity and the other of lower affinity.[119,120] Both are seven transmembrane spanning, G-protein-linked receptors of the rodopsin superfamily. While the A receptor binds to IL-8, the B receptor binds IL-8 as well as NAP-2 and GROα. As mentioned above, neutrophils are the primary targets for IL-8, and respond to this mediator in vitro by chemotaxis,[121] release of granule enzymes,[116,122] respiratory burst,[123] up-regulation of CR1 and CD11/CD18 expression on the surface, and increased adherence to unstimulated endothelial cells,[115] as well as by activation of intracellular signals, such as a transient rise in free Ca^{2+} and elevation of cAMP levels.[123,124] IL-8, in addition, mediates the transmigration of neutrophils across endothelium,[125] has chemotactic activities for T lymphocytes and basophils,[114,115] although much less effectively than for neutrophils, and is also an angiogenic factor.[126] IL-8, when administered intravenously, causes neutrophilia, plasma exudation and neutrophil infiltration into tissues which is maintained for some hours when administered locally, and granulocytosis upon intravenous injection in rabbits.[114,115,127,128] IL-8 and several other α-chemokines have been shown to be produced in various disease states, including rheumatoid arthritis, inflammatory bowel disease, atherosclerosis, asthma, tuberculosis, psoriasis, and a variety of respiratory syndromes. The generation of antagonists specific for IL-8 and other α-chemokines or chemokine receptor specific antibodies may provide a means of controlling both acute and chronic inflammatory disease states.[114,115]

INTERLEUKIN-10 (IL-10)

IL-10 is a prototypical anti-inflammatory cytokine, whose physiologic function is to control a T helper type 1 (Th1) cell-mediated inflammation. IL-10 is a 18 kDa polypeptide with a high degree of homology with an open reading frame (BCRF1) in the Epstein Barr virus (EBV) genome.[129] It has been postulated that the virus may have incorporated a cellular IL-10 gene as a mechanism for inhibiting antiviral defense and promoting its survival. IL-10 is secreted by monocytes/macrophages, B lymphocytes, keratinocytes, but its major sources are some subclasses of CD4+ T lymphocytes, primarily the Th0 and Th2 types.[130] IL-10 blocks activation of proinflammatory cytokine synthesis by many cells, including Th1 cells[131] and NK cells,[132] and can act as a co-stimulator of the growth of B cells, thymocytes and mast cells.[130,133] In monocytes/macrophages, IL-10 down-regulates constitutive and IFNγ-induced class II MHC expression,[134] ROI and nitric oxide production,[135] and suppresses the synthesis of proinflammatory cytokines.[134,136] Basically, IL-10 inhibits the accessory functions of macrophages.[137] In murine systems, IL-10, inhibits the development of Th1 effector cells from naive CD4+ T lymphocytes, and induces a Th2 dominant population.[138] This effect is largely due to inhibition of IL-12 production by macrophages, IL-12 being an important Th1-inducing cytokine.[139] Furthermore, because of its capacity to inhibit the production and release of TNFα and other proinflammatory cytokines by macrophages in response to LPS, IL-10 reduces the lethality of experimental septic shock in mice.[140-142]

The potential therapeutic utility of this cytokine is therefore caused by excessive macrophage and T cell-mediated inflammatory reactions (endotoxic shock, autoimmune diseases, transplant rejections and chronic inflammations) in various disorders.

INTERLEUKIN-12 (IL-12)

NK cell stimulatory factor or IL-12 is a heterodimeric factor of 70 kDa (p70), formed by two covalently-linked glycosylated chains of approximately 40 kDa (p40) and 35 kDa (p35), whose co-expression is necessary for biologic activity.[143] Although the molecular cloning of IL-12 was published only few years ago, its physiologic importance is now well established, as it plays a central role in macrophage-T cell interactions and protective cell mediated immunity against intracellular microbes.[139] IL-12 is produced mainly by mononuclear phagocytic cells, B cells, neutrophils and other types of antigen presenting cells (APC).[144] The most efficient inducers of IL-12 production from phagocytic cells are bacteria, bacterial products and intracellular parasites.[139] The ability of phagocytic cells to produce IL-12 is regulated by several cytokines with activating or suppressing effects on the producer cells; IFNγ and GM-CSF enhance the production of IL-12, whereas IL-4, IL-10, IL-13, and TGFβ inhibit IL-12 production.[145-148] IL-12 triggers several biologic activities of NK cells and T lymphocytes, on which it induces the production of lymphokines, particularly IFNγ, enhancement of cell-mediated cytotoxicity, and mitogenic effects.[139] In addition, IL-12 is required for optimal differentiation of cytotoxic T lymphocytes[149] and appears to have an obligatory role as inducer of Th1 differentiation and suppression of IgE synthesis.[139,150] The early decision toward Th1 or Th2 cells in the immune response is dependent on the balance between IL-12, which favors Th1 responses, and IL-4, which favors Th2 responses.[150] IL-12 plays a pivotal role in cell mediated inflammatory reactions, and in vivo IL-12 appears to be critical both for initiating and sustaining a cellular immune response.[151,152] The relationship of IL-12 with other cytokines and immunoregulatory components is being rapidly established.[139] The production of IL-12 and IFNγ early in bacterial and parasitic infections represents an important mechanism of the innate resistance to infection and its role has been clearly demonstrated in infection with intracellular pathogens such as *L.monocytogenes*,[152,153] *T.gondii*,[151,154] *L.major*,[155] *C.albicans*,[156] *M.leprae* and *M.tuberculosis*.[157] Furthermore, the profound depression of IL-12 production observed in human immunodeficiency virus (HIV) individuals may, for example, contribute to the immunodeficiency observed in acquired immunodeficiency syndrome (AIDS) patients.[158] Therefore, IL-12 is considered a potential candidate for therapeutic use in diverse infectious diseases including AIDS, in possible vaccines against these diseases, and also in solid tumors.[139]

GROWTH RELATED GENE PRODUCT-α (GROα)/MELANOMA GROWTH-STIMULATORY ACTIVITY (MSGA)

GROα is a 73 amino acid, 8 kDa protein that is highly chemotactic for PMN.[159-161] GROα was originally identified as an endogenous growth factor for human melanoma because of its properties as an autocrine growth factor for melanoma cells (melanoma growth stimulatory activity, MGSA).[114,115] MGSA/GROα has also been known as NAP-3. GRO has two additional variants, GROβ and GROγ.[114,115] They are probably homologues of murine macrophage inflammatory peptide-2α (MIP-2α) and MIP-2β, respectively. Because of their structural relationship to IL-8 (30% homology), GROs are included in the α-chemokine superfamily.[114,115] The three GRO genes have approximately 90% sequence identity[162] and share similar PMN-activating properties.[114,115] GRO peptides have been shown to be produced by a variety of normal cells, including epithelial and endothelial cells, fibroblasts, monocytes, neutrophils, and tumor cells.[115] Like IL-8, GROα appears to be an inflammatory mediator, since it has powerful chemotactic and activatory properties on PMN, including degranulation, increased expression of adhesion molecules and in vivo recruitment of neutrophils to sites of injection.[159-161] Elevated levels of GRO can also be detected in biopsies and extracted scales of psoriatic lesions, suggesting that it plays a role in the pathogenesis of inflammatory diseases, and several studies have also demonstrated hyperexpression of GRO in neoplastic cells.[114,115] Recently, GROα and GROβ, but not GROγ, have been found to specifically inhibit growth factor-stimulated proliferation of capillary endothelial cells.[163] Furthermore, GROβ inhibited Lewis lung tumor growth in mice by suppression of tumor-induced neovascularization.[163]

MACROPHAGE INFLAMMATORY PROTEIN-1α AND -1β (MIP-1α/β)

The macrophage inflammatory proteins (MIP)-1α and MIP-1β are members of the β-chemokine (C-C subfamily) family, which also includes MCP-2, MCP-3, I-309/TCA3, and RANTES.[114,115] MIP-1α (previously known as LD-78) and MIP-1β (previously known as ACT-2) were characterized in the mouse as a single small (8-12 kDa) protein species purified from a stimulated macrophage cell line.[164] MIP-1α and MIP-1β in mouse and man are both about 70% identical to each other in their mature secreted forms (68-69 amino acids). Data from a number of laboratories suggest that both MIP-1α and MIP-1β are expressed primarily in T cells, B cells, monocytes, neutrophils, and Langerhans cells stimulated with antigens, specific agonists, or mitogens.[114,115] MIP-1α acts as potent chemotactic/activating factor for monocytes and subpopulations of T and B lymphocytes, eosinophils, and also activates several effector functions of macrophages and neutrophils,[114,115] such as the generation of hydrogen peroxide,[164] and the secretion of TNFα, IL-1α and IL-6.[165] MIP-1α may be pyrogenic[166] and is also implicated in negative regulation of myelopoiesis, since it has suppressive activity on hematopoietic stem cells and early subsets of hematopoietic progenitor cells, both in vitro[167] and in vivo.[168] MIP-1β is closely related to MIP-1α, with which it may function in concert or, in some instances, in antagonism.[165] Recently, a preferential migration of activated CD4+ and CD8+ T cells in response to MIP-1α and MIP-1β has been reported.[169]

MONOCYTE CHEMOTACTIC PROTEINS (MCP-1, MCP-2 AND MCP-3)

Monocyte chemotactic protein-1 (MCP-1) is a glycoprotein with a molecular weight of 9-15 kDa and is another cytokine belonging to the family of the so-called β-chemokines, as MIP-1α/β.[114,115] MCP-1 has also been known as monocyte chemoattractant and activating factor (MCAF). MCP-1 bears a high degree of homology with the murine JE gene, which was originally identified in mitogen-stimulated non-neoplastic fibroblasts.[170] Almost all cells or tissues examined make MCP-1 upon stimulation with a variety of agents, but the targets are limited to monocytes and basophils.[114,115] MCP-1/JE is known to strongly stimulate chemotaxis of peripheral blood monocytes, and to modulate several monocyte responses, including the respiratory burst, calcium influx,[171] adhesion molecule expression[1,72] release of lysosomal enzymes, and monocyte mediated inhibition of tumor cell growth.[173] MCP-1 also attracts basophils, eosinophils, lymphocytes but not neutrophils.[114,115] In addition, MCP-1 directly upregulates monocyte cytokine production[174] and cytostatic activity.[175] Interestingly, monocytes are the primary source of such cytokines, and respond to TNFα with increased MCP-1 production in an autocrine manner.[176] MCP-1 has been detected by in situ hybridization or immunohistochemistry in the lesions of rheumatoid arthritis, atherosclerosis, idiopathic pulmonary fibrosis and many tumors.[114,115] Recent reports have revealed the existence of MCP-2 and MCP-3, with, respectively, 62% and 73% amino acid identity to MCP-1.[177] MCP-2 and MCP-3 share the chemoattractant specificity of MCP-1 for monocytes in vivo.[177]

INTERFERON INDUCIBLE PROTEIN-10 (IP-10)

IP-10 is another member of the α-chemokine family, and was originally identified as an IFNγ-inducible protein of 10 kDa from the human U937 monocytic leukemia cell line.[178] IP-10 can be produced by monocytes, T lymphocytes, neutrophils, keratinocytes, fibrobalsts, and other cell types stimulated with IFNγ, alone or in combination with IL-2, LPS or IFNβ.[179] IP-10 is the only α-chemokine member that chemoattracts monocytes and lymphocytes rather than neutrophils.[180] In fact, IP-10 is devoid of the -E-L-R- amino acid sequence preceding the first two cysteine residues, whose conservation is essential to confer neutrophil chemoattractant properties to α-chemokines.[180,181] IP-10 can readily be detected in cells participating in delayed-type hypersensitivity responses and in lesions of lepromatous leprosy patients.[182] Interestingly, psoriatic plaques which are characterized by neutrophil infiltration are also a rich source of IP-10,[183] suggesting that this protein may play a role in the development of chronic inflammatory responses.

STEM CELL FACTOR (SCF)

Stem cell factor (SCF), also known as Kit ligand (KL), steel factor (SLF) or mast cell growth factor (MGF), is a 30 kDa glycoprotein involved in the development of hematopoietic, gonadal and pigment cell lineages.[184,185] It is secreted by bone marrow stromal cells, fibroblasts and endothelial cells and in addition to neutrophils, the expression of SCF mRNA has been detected in brain, lung, heart and kidney of developing mouse embryo.[186] The expression of SCF is usually downregulated by TNFα, IL-1 and TGFβ1 in vitro.[186] SCF has a very wide range of activities with direct effects on myeloid and lymphoid cell development and powerful synergistic effects with other growth factors such as GM-CSF, erythropoietin and IL-7.[185] Additionally, SCF enhances IL-2 to stimulate the proliferation and cytotoxic activity of human NK subsets constitutively expressing the high affinity IL-2R and *c-kit*.[187] In mice, mutations of the W locus which encodes *c-kit* lead to changes in coat color, anemia and defective gonad development. Alternative mRNA splicing gives rise to two forms of biologically active SCF, both of which have a transmembrane domain and are inserted into the cell membrane.[188] The larger form contains a peptide cleavage site and is processed to yield secreted SCF.[186] SCF has already been shown to possess potential clinical utility in enhancing the ability of CSF to stimulate the production of hematopoietic cells, and to mobilize hematopoietic progenitor cells to peripheral blood.[189]

COLONY STIMULATING FACTORS (CSF)

The name colony stimulating factor (CSF), although only an operational definition, is conceptually useful since this group of cytokines functions mainly to regulate proliferation and differentiation of various types of hematopoietic cells (e.g., granulocytes, macrophages and erythroid cells). There is functional overlap between CSFs and some interleukins. For example IL-3 was formerly called multi-CSF among other names, because it promotes growth and differentiation of both myeloid and erythroid precursors.[190] Thus its activities overlap those of granulocyte macrophage-CSF (GM-CSF).

GM-CSF supports proliferation of bone-marrow derived progenitors of both granulocytes and macrophages, but at a later stage of differentiation than those cells targeted by IL-3.[191] GM-CSF is produced by activated T lymphocytes and also by fibroblasts, endothelial cells, monocytes and stromal cells in the bone marrow.[192] Synthesis and secretion of GM-CSF are regulated, for example, by IL-1 or IFNγ, and often occur in response to antigenic stimulation. Thus, GM-CSF probably has a major role in host defense mechanisms requiring increased numbers of granulocytes and macrophages and this could be more important than the part it plays in normal hematopoiesis. GM-CSF also exerts effects on mature granulocytes (including eosinophils and basophils) and macrophages. For example, it stimulates secretion of soluble inflammatory mediators and cytokines, it increases phagocytosis and ADCC, it primes for enhanced oxidative burst to primary stimuli, and it enhances bactericidal and parasite killing activities.[191] In vivo, GM-CSF administration results in an increase in circulating neutrophils, eosinophils, monocytes, and all progenitor cells, and increased numbers and activation status of tissue macrophage.

Like GM-CSF, granulocyte (G)-CSF, acts on both granulocyte precursors and the mature cells and is produced by bone marrow stromal cells, endothelial cells, and fibroblasts in response to specific stimuli.[193] As well as enhancing proliferation of granulocyte precursors in bone marrow, G-CSF activates neutrophils in several ways, stimulating phagocytosis and antibody-mediated cytotoxicity against targets such as tumor cells, and priming them for superoxide anion production.[193] G-CSF is widely used clinically in the treatment of patients with neutropenia after cancer chemotherapy.

Just as granulocytes have a specific CSF to regulate their activities, macrophages, too, are controlled by a relatively specific cytokine, macrophage (M)-CSF, also formerly called CSF-1.[194] M-CSF alone stimulates survival, proliferation and differentiation of mononuclear phagocytic cells from the undifferentiated but committed bone marrow precursor cell to the mature, non-dividing macrophage.[194] Monocytes and macrophages are the main sources of M-CSF in normal tissues, suggesting that M-CSF functions in an autocrine manner in vivo. However, a variety of different cell types, including bone marrow stromal cells, fibroblasts, endothelial cells, osteoblasts, keratinocytes, astrocytes and uterine epithelial cells can synthesize M-CSF. Other than having a role in controlling macrophage development, M-CSF stimulates monocyte/macrophages to secrete IL-1 and prostaglandin E, to potentiate synthesis of peroxides, and to activate phagocytosis and cell killing.[194] M-CSF is also involved in the regulation of cells of the female reproductive tract during pregnancy, osteoclast progenitor cell differentiation and microglial cell proliferation.[194]

REFERENCES

1. Sen GC, Lengyel P. The interferon system. A bird's eye view of its biochemistry. J Biol Chem 1992; 267:5017-5020.
2. Pestka S, Langer JA, Zoon KC et al. Interferon and their action. Ann Rev Biochem 1987; 56:727-777.
3. Uze G, Lutfalla G, Gressler I. Genetic transfer of a functional human interferon-alpha receptor into mouse cells: cloning and expression of its cDNA. Cell 1990; 60:225-234.
4. Novick D, Cohen B, Rubinstein M. The human interferon α/β receptor: characterization and molecular cloning. Cell 1994; 77:391-400.
5. Heron I, Hokland M, Berg K. Enhanced expression of β2-microglobulin and HLA antigens on human lymphoid cells by IFN. Proc Natl Acad Sci USA 1978; 75:6215- 6219.
6. Trinchieri G, Santoli D. Antiviral activity induced by culturing lymphocytes with tumor derived or virus transformed cells. Enhancement of human natural killer cell activity by interferon and antagonistic inhibition of susceptibility of target cells to lysis. J Exp Med 1978; 147:1314-1333.
7. Grossber SE. Interferons: an overview of their biological and biochemical

properties. In: Pfeffer LM, ed. Mechanisms of Interferon Action. Boca Raton, Fla: CRC Press, 1987:1-32.

8. Djeu JY, Blanchard DK. Regulation of human polymorphonuclear neutrophil (PMN) activity against *Candida albicans* by large granular lymphocytes via release of a PMN-activating factor. J Immunol 1987; 139:2761-2767.

9. Dianzani F. Interferon treatments: how to use an endogenous system as a therapeutic agent. J Interf Res 1992; 12:109-118.

10. Fitzgerals-Bocarsly P. Human natural interferon producing cells. Pharmac Ther 1993; 60:39-62.

11. Vilcek J, Gray PW, Rinderknecht E et al. Interferon-γ, a lymphokine for all seasons. Lymphokines 1985; 11:1-32.

12. Farrar MA, Schreiber RD. The molecular cell biology of Interferon-γ and its receptor. Ann Rev Immunol 1993; 11:571-611.

13. Schindler C, Darnell JE Jr. Transcriptional responses to polypeptide ligands: the JAK-STAT pathway. Annu Rev Biochem 1995; 64:621-651.

14. Berton G, Cassatella MA. Modulation of neutrophil functions by interferon gamma. In: Coffey RG, ed. Granulocytes Responses to Cytokine: Basic and Clinical Researches. New York: Marcel Dekker Inc., 1992: 437-456.

15. Perussia B, Dayton F, Lazarus S et al. Immune interferon induces the receptor for monomeric IgG1 on human monocytic and myeloid cells. J Exp Med 1983; 158:1092-1113.

16. Basham TY, Smith WK, Merigan TC. Interferon enhances antibody-dependent cellular cytotoxicity when suboptimal concentrations of antibody are used. Cell Immunol 1984; 88:393-400.

17. Berton G, Zeni L, Cassatella MA et al. Gamma interferon is able to enhance the oxidative metabolism of human neutrophils. Biochem Byophis Res Commun 1986; 138:1276-1282.

18. Cassatella MA, Cappelli R, Della Bianca V et al. Interferon gamma activates neutrophil oxygen metabolism and exocytosis. Immunology 1988; 63:499-506.

19. Ezekowitz RAB, and the international chronic granulomatous disease cooperative study group. A controlled trial of interferon gamma to prevent infection in chronic granulomatous disease. N Engl J Med 1991; 324:509-516.

20. Wahl SM. Transforming growth factor β: the good, the bad and the ugly. J Exp Med 1994; 180:1587-1590.

21. McCartney-Francis NL, Wahl SM. Transforming growth factor-β: a matter of life and death. J Leuk Biol 1994; 55:401-409.

22. Shull MM, Ormsby I, Kier AB et al. Targeted disruption of the mouse transforming growth factor β1 gene results in multifocal inflammatory disease. Nature 1992; 359:693-699.

23. Kulkarni AB, Huh CH, Becker D et al. Transforming growth factor β1 null mutation in mice causes excessive inflammatory response and early death. Proc Natl Acad Sci Usa 1993; 90:770-774.

24. Myazono K, Hellman U, Wernstedt C et al. Latent high molecular weight complex of TGFβ1. J Biol Chem 1988; 263:6407-6415.
25. Wahl SM, Hunt DA, Wakefield L et al. Transforming growth factor beta induces monocyte chemotaxis and growth factor production. Proc Natl Acad Sci USA 1987; 84:5788-5792.
26. Brandes ME, Mai UE, Ohura K et al. Type-I transforming growth factor beta receptors mediate chemotaxis to transforming growth factor beta. J Immunol 1991; 147:1600-1606.
27. Massague J. The transforming growth factor-β family. Annu Rev Cell Biol 1990; 6:597-641.
28. Tsunawaki S, Soprn M, Ding A, Nathan C. Deactivation of macrophages by Transforming growth factor-β. Nature 1998; 334:260-262.
29. Chantry D, Turner M, Abney E et al. Modulation of cytokine production by TGFβ. J Immunol 1989; 142:4295-4300.
30. Czarniecki CW, Chiu HH, Wong GHW et al. TGFβ1 modulates the expression of class II histocompatibilty antigens on human cells. J Immunol 1988; 140:4217-4223.
31. Welch GR, Wong ML, Wahl SM. Selective induction of FcγRIII on human monocytes by TGFβ. J Immunol 1990; 144:3444-3448.
32. Vodovotz Y, Bogdan C, Paik J et al. Mechanisms of suppression of macrophage nitric oxide release by transforming growth factor-β. J Exp Med 1993; 178:605-613.
33. Border AW, Ruoshlati E. Transforming growth factor-β in disease: the dark side of the tissue repair. J Clin Invest 1992; 90:1-7.
34. Sporn MB, Roberts AB. A major advance in the use of growth factors to enhance wound healing. J Clin Invest 1993; 92:2565-2566.
35. Border WA, Noble NA. Transforming growth factor-β in tissue fibrosis. N Engl J Med 1994; 331:1286-1292.
36. Reibman J, Meixler S, Lee TC et al. Transforming growth factor beta1, a potent chemoattractant for human neutrophils, bypasses classic signal-transduction pathways. Proc Natl Acad Sci USA 1991; 88:6805-6809.
37. Parekh, T, Saxena B, Reibman J et al. Neutrophil chemotaxis in response to TGFβ isoforms (TGFβ1, TGFβ2, TGFβ3) is mediated by fibronectin. J Immunol 1994; 152:2456-2466.
38. Bank U, Reinhold D, Kunz D et al. Effects of interleukin-6 and transforming growth factor beta on neutrophil elastase release. Inflammation 1995; 19:83-99.
39. Tracey, KJ. Tumor necrosis factor-alpha. In: Thomson A, ed. The cytokine handbook. 2nd edition, New York: Academic Press, 1994:289-304.
40. Gearing AJ, Beckett P, Christodolou M et al. Processing of tumor necrosis factor alpha precursor by metalloproteinases. Nature 1994; 370:555-557.
41. Smith CA, Farrah T, Goodwin RG. The TNF receptor superfamily of cellular and viral proteins: activation, costimulation and death. Cell 1994; 76:959-962.
42. Brockhaus M, Schonfeld H-J, Schlaeger E-J et al. Identification of two types of tumor necrosis factor receptors on human cell lines by monoclonal antibodies. Proc Natl Acad Sci USA 1990; 87:3127-3131.

43. Carswell EA, Old LJ, Kassel RL et al. An endotoxin-induced serum factor that causes necrosis of tumors. Proc Natl Acad Sci USA 1995; 72:3666-3670.
44. Vassalli P. The pathophysiology of tumor necrosis factors. Ann Rev Immunol 1992; 10:411-452.
45. In: Beutler D, ed. Tumor Necrosis Factors: the molecules and their emerging role in medicine. New York: Raven Press, 1992:1-590.
46. Pfeffer K, Matsuyama T, Kundig TM et al. Mice deficient of the 55 kd tumor necrosis factor receptor are resistant to septic shock, yet succumb to *L.monocytogenes* infection. Cell 1993; 73:457-467.
47. Elliott MJ, Maini RN, Feldmann M et al. Randomised double-blind comparison of chimeric monoclonal antibody to tumor necrosis factor-α (cA2) versus placebo in rheumatoid arthritis. Lancet 1994; 344:1105-1110.
48. Derkx B, Taminiau J, Radema S et al. Tumor necrosis factor antibody treatment in Crohn's disease. Lancet 1994; 342:173-174.
49. Pober JS, Gimbrone MA, Lapierre LA et al. Overlapping patterns of activation of human endothelial cells by interleukin-1, tumor necrosis factor, and immune interferon. J Immunol 1986; 137:1893-1896.
50. Gamble JR, Harlan JM, Klebanoff SJ et al. Stimulation of the adherence of neutrophils to umbilical vein endothelium by human recombinant tumor necrosis factor. Proc Natl Acad Sci USA 1985; 82:8667-8671.
51. Della Bianca V, Dusi S, Nadalini KA et al. Role of 55 and 75 kDa TNF receptors in the potentiation of Fc-mediated phagocytosis in human neutrophils. Biochem Biophys Res Commun 1995; 214:44-50.
52. Klebanoff SJ, Vadas MA, Harlan JM et al. Stimulation of neutrophils by tumor necrosis factor. J Immunol 1986; 136:4220-4225.
53. Shalaby MR, Aggarwal BB, Rinderknecht E et al. Activation of human polymorphonuclear neutrophil functions by IFNγ and TNFs. J Immunol 1985; 135:2069-2073.
54. Richter J, Andersson T, Olsson I. Effect of TNFα and granulocyte-macrophage colony-stimulating factor on neutrophil degranulation. J Immunol 1989; 142:3199-3205.
55. Nathan CF. Neutrophil activation on biological surfaces. J Clin Invest 1987; 80:1550-1560.
56. Dinarello CA. The interleukin-1 family: 10 years of discovery. FASEB J 1994; 8:1314-1324.
57. Arend WP. Interleukin-1 receptor antagonist. Adv Immunol 1993; 54:167-227.
58. Colotta F, Dower SK, Sims JE et al. The type II "decoy" receptor: novel regulatory pathway for interleukin-1. Immunol Today 1994; 15:562-566.
59. Cerretti DP, Kozlosky CJ, Mosley B et al. Molecular cloning of the IL-1β converting enzyme. Science 1992; 256:97-100.
60. Colotta F, Re F, Polentarutti N et al. Modulation of granulocyte survival and programmed cell death by cytokines and bacterial products. Blood 1992; 80:2012-2020.

61. Mantovani A, Bussolino F, Dejana E. Cytokine regulation of endothelial cell function. FASEB J 1992; 6:2591-2599.
62. Sapolsky R, Rivier C, Yamamoto G et al. Interleukin-1 stimulates the secretion of hypothalamic corticotropin-releasing factor. Science 1987; 238:522-524.
63. Dinarello CA. Interleukin-1 and the pathogenesis of the acute-phase response. N Engl J Med 1984; 311:1413-1418.
64. Bendtzen K, Mandrup-Poulsen T, Nerup J et al. Cytotoxicity of human interleukin 1 for pancreatic islets of Langerhans. Science 1986; 232:1545-1547.
65. Arend WP, Joslin FG, Massoni RJ. Effects of immune complexes on production by human monocytes of interleukin 1 or an interleukin 1 inhibitor. J Immunol 1985; 134:3868-3875.
66. Seckinger PJ, Lowenthal W, Williamson K et al. A urine inhibitor of interleukin 1 activity that blocks ligand binding. J Immunol 1987; 139:1541-1545.
67. Balavoine JF, de Rochemonteix B, Williamson K et al. Prostaglandin E_2 and collagenase production by fibroblasts and synovial cells is regulated by urine-derived human interleukin 1 and inhibitor(s). J Clin Invest 1986; 78:1120-1124.
68. Dripps DJ, Brandhuber BJ, Thompson RC et al. Interleukin-1 (IL-1) receptor antagonist binds to the 80 kDa IL-1 receptor but does not initiate IL-1 signal transduction. J Biol Chem 1991; 266:10331-10336.
69. Hannum CH, Wilcox CJ, Arend WP et al. Interleukin-1 receptor antagonist activity of a human interleukin-1 inhibitor. Nature 1990; 343:336-340.
70. Haskill S, Martin G, Van Le L et al. cDNA cloning of an intracellular form of the human interleukin-1 receptor antagonist associated with epithelium. Proc Natl Acad Sci USA 1991; 88:3681-3685.
71. Muzio M, Re F, Sironi M et al. Interleukin-13 induces the production of Interleukin-1 receptor antagonist (IL-1ra) and the expression of the mRNA for the intracellular (keratinocyte) form of IL-1ra in human myelomonocytic cells. Blood 1994, 83:1738-1743.
72. Muzio M, Polentarutti N, Sironi M et al. Cloning and characterization of a new isoform of the interleukin-1 receptor antagonist. J Exp Med 1995: 182:623-628.
73. Ozaki Y, Ohashi T, Kume S. Potentiation of neutrophil function by recombinant DNA-produced interleukin-1α. J Leuk Biol 1987; 42:621-627.
74. Sullivan GW, Carper HT, Sullivan JA et al. Both recombinant interleukin-1 (beta) and purified human monocyte interleukin-1 prime human neutrophils for increased oxidative activity and promote neutrophil spreading. J Leuk Biol 1989; 45:389-395.
75. Dularay B, Elson CJ, Clements-Jewery S et al. Recombinant human interleukin-1β primes human polymorphonuclear leukocytes for stimulus-induced myeloperoxidase release. J Leuk Biol 1990; 47:158-163.

76. Ferrante A, Nandoskar M, Walz A et al. Effects of tumor necrosis factor-alpha and interleukin 1 alpha and beta on human neutrophil migration, respiratory burst and degranulation. Int Arch Allergy Appl Immunol 1988; 86:82-91.
77. Yagisawa M, Yuo A, Kitagawa S et al. Stimulation and priming of human neutrophils by IL-1α and IL-1α: complete inhibition by IL-1 receptor antagonist, and no interaction with other cytokines. Exp Hematol 1995; 23:603-608.
78. Goldsmith MA, Greene WC. Interleukin-2. In: Nicola N, ed. Guidebook to Cytokine and their Receptor. Oxford: Oxford Univ Press, Sambrook and Tooze publication, 1994:27-30.
79. Paul WE. Interleukin-4: a prototypic immunoregulatory lymphokine. Blood 1991; 77:1859-1870.
80. Kishimoto T, Taga T, Akira S. Cytokine signal transduction. Cell 1994; 76:253-262.
81. Finkelman FD, Holmes J, Katona IM et al. Lymphokine control of in vivo immunoglobulin isotype selection. Annu Rev Immunol 1990; 8:303-333.
82. Seder RA, Paul WE. Acquisition of lymphokine-producing phenotypes by CD4+ T cells. Ann Rev Immunol 1994; 12:635-673.
83. Kuhn R, Rajewsky K, Muller W. Generation and analysis of interleukin-4 deficient mice. Science 1991; 254:707-710.
84. Sadick MD, Heinzel FP, Holaday BJ et al. Cure of murine leishmaniasis with anti-interleukin 4 monoclonal antibody. Evidence for a T cell-dependent, interferon gamma-independent mechanism. J Exp Med 1990; 171:115-127.
85. Finkelman FD, Urban JF, Paul WP et al. In: Spits H, ed. IL-4: Structure and Function. Boca Raton: CRC Press, 1992:33-54.
86. Tepper RI, Coffman RL, Leder P. An eosinophil-dependent mechanism for the antitumor effect of interleukin-4. Science 1992; 257:548-551.
87. Stuart PM, Zlotnik A, Woodward JG. Induction of class I and II MHC antigen expression on murine bone marrow-derived macrophages by IL-4 (B cell stimulatory factor 1). J Immunol 1988; 140:1542-1547.
88. Crawford RM, Flinboom DS, Ohara J et al. B cell stimulatory factor-1 (interleukin-4) activates macrophages for increased tumoricidal activity and expression of Ia antigens. J Immunol 1987; 139:135-141.
89. Lehn M, Weisner WY, Enghelorn S et al. IL-4 inhibits H_2O_2 production and antileishmanial capacity of human cultured monocytes mediated by IFNγ. J Immunol 1989; 143:3020-3024.
90. Hart PH, Vitti GF, Burgess DR et al. Potential anti-inflammatory effects of interleukin-4: suppression of human monocyte tumor necrosis factor-α, interleukin-1, and prostglandin E_2. Proc Natl Acad Sci USA 1989; 86:3803-3807.
91. Boey H, Rosenbaum R, Castracane J, Borish L. Interleukin-4 is a neutrophil activator. J Allergy Clin Immunol 1989; 83:978-984.

92. Bober LA, Waters TA, Pugliese Sivo CC et al. IL-4 induces neutrophilic maturation of HL-60 cells and activation of human peripheral blood neutrophils. Clin Exp Immunol 1995; 99:129-136.
93. Zurawzki G, de Vries JE. Interleukin 13 elicits a subset of the activities of its close relative interleukin 4. Stem Cell 1994; 12:169-174.
94. Welham MJ, Learmonth L, Bone H et al. Interleukin-13 signal transduction in lymphohemopoietic cells. Similarities and differences in signal transduction with interleukin-4 and insulin. J Biol Chem 1995; 270: 12286-12296.
95. Obiri NI, Debinski W, Leonard WJ et al. Receptor for interleukin 13. Interaction with interleukin 4 by a mechanism that does not involve the common gamma chain shared receptors for interleukin 2,4,7,9,and 15. J Biol Chem 1995; 270:8797-8804.
96. Kishimoto T, Akira S, Taga T. Interleukin-6 and its receptor: a paradigm for cytokines. Science 1992; 258: 593-597.
97. Akira S, Taga T, Kishimoto T. Interleukin-6 in biology and medicine. Adv Immunol 1993; 54:1-78.
98. Taga T, Kishimoto T. Cytokine receptors and signal transduction. FASEB J 1992; 6:3387-3396.
99. Van Damme J, Opdenakker G, Simpson RJ et al. Identification of the human 26-kD protein, interferon β2 (IFNβ2) as a B cell hybridoma/plasmocytoma growth factor induced by IL-1 and tumor necrosis factor. J Exp Med 1987; 165:914-919.
100. Poli V, Balena R, Fattori E et al. Interleukin-6 deficient mice are protected from bone loss caused by estrogen depletion. EMBO J 1994; 13:1189-1196.
101. Jilka RL, Hangoc G, Girasole G et al. Increased osteoclast development after estrogen loss: mediation by interleukin-6. Science 1992; 257:88-91.
102. Suematsu S, Matsusaka T, Matsuda T et al. Generation of plasmacytomas with the chromosomal translocation t(12;15) in interleukin-6 transgenic mice. Proc Natl Acad Sci USA 1992; 89:232-235.
103. Fong Y, Tracey KJ, Moldawer LL et al. Antibodies to cachectin/tumor necrosis factor reduce interleukin 1β and interleukin 6 appearance during lethal bacteremia. J Exp Med 1989; 170:1627-1633.
104. Waage A, Brandzaeg P, Halstensen A et al. The complex pattern of cytokines in serum from patients with meningococcal septic shock. Association between interleukin 6, interleukin 1, and fatal outcome. J Exp Med 1989; 169:333-338.
105. Kopf M, Baumann H, Freer G, Freudenberg M et al. Impaired immune and acute-phase responses in interleukin-6-deficient mice. Nature 1994; 368:339-342.
106. Fattori E, Cappelletti M, Costa P et al. Defective inflammatory response in inteleukin-6 deficient mice. J Exp Med 1994; 180:1243-1250.
107. Kapp A, Zeck-Kapp G. Activation of the oxidative metabolism in human polymorphonuclear neutrophilic granulocytes: the role of immuno-stimulating cytokines. J Invest Dermatol 1990; 95:94S-99S.

108. Henschler R, Lindemann A, Brach MA et al. Expression of functional receptors for interleukin-6 by human polymorphonuclear leukocytes. FEBS Lett 1991; 283:47-51.
109. Brom J, Konig W. Cytokine-induced (interleukin-3,-6,-8 and tumor necrosis factor-beta) activation and deactivation of human neutrophils. Immunology 1992; 75:281-285.
110. Borish L, Rosenbaum R, Albury L et al. Activation of neutrophils by recombinant interleukin-6. Cell Immunol 1989; 121:280-289.
111. Taub DD, Oppenheim JJ. Review of the chemokine meeting. The third international symposium of chemotactic cytokines. Cytokine 1993; 5:175-179.
112. Kelner GS, Kennedy J, Bacon KB et al. Lymphotactin: a cytokine that represents a new class of chemokine. Science 1994: 266:1395-1399.
113. Kennedy J, Kelner JS, Kleyensteuber S et al. Molecular cloning and functional characterization of human lymphotactin. J Immunol 1995; 155:203-209.
114. Oppenheim JJ, Zacharie COC, Mukaida N et al. Properties of the novel proinflammatory supergene "intercrine" cytokine family. Ann Rev Immunol 1991; 9:617-648.
115. Baggiolini M, Dewald B, Moser B. Interleukin-8 and related chemotactic cytokines-CXC and CC chemokines. Adv Immunol 1994; 55:97-179.
116. Schroder JM, Mrowietz U, Morita E et al. Purification and partial biochemical characterization of a human monocyte-derived neutrophil activating peptide that lacks interleukin-1 activity. J Immunol 1987; 139:3474-3483.
117. Yoshimura TK, Matsushima K, Tanaka S et al. Purification and partial biochemical characterization of a human monocyte-derived neutrophil chemotactic factor that has peptide sequence similarity to other host defense cytokines. Proc Natl Acad Sci USA 1987; 84:9233-9237.
118. Hèbert CA, Luscinskas FW, Kiely J-M et al. Endothelial and leukocyte forms of IL-8. J Immunol 1990; 145:3033-3040.
119. Holmes WE, Lee J, Kuang WJ et al. Structural and functional expression of a human interleukin-8 receptor. Science 1991; 253:1278-1280.
120. Murphy PM, Tiffany HL. Cloning of complementary DNA encoding a functional human Interleukin-8 receptor. Science 1991; 253:1280-1283.
121. Van Damme J, van Beeumen J, Opdenakker G et al. A novel NH_2-terminal sequence-characterized human monokine possessing neutrophil chemotactic, skin-reactive, and granulocytosis-promoting activity. J Exp Med 1988; 167:1364-1376.
122. Walz A, Peveri P, Ashauer H et al. Purification and amino acid sequencing of NAF, a novel neutrophil-activating factor produced by monocytes. Biochem Biophys Res Commun 1987; 149:755-761.
123. Walz A, Meloni F, Clark-Lewis I et al. $[Ca^{2+}]i$ changes and respiratory burst in human neutrophils and monocytes induced by NAP-1/interleukin-8, NAP-2, and GRO/MGSA. J Leuk Biol 1991; 50:279-286.

124. Brandt E, Petersen F, Flad H-D. Recombinant tumor necrosis factor-α potentiates neutrophil degranulation in response to host defense cytokines neutrophil-activating peptide-2 and IL-8 by modulating intracellular cyclic AMP levels. J Immunol 1992; 149:1356-1364.
125. Huber AR, Kunkel SL, Todd III RF et al. Regulation of transendothelial migration by endogenous interleukin-8. Science 1991; 254:99-102.
126. Koch AE, Kunkel SL, Harlow LA et al. IL-8 as macrophage-derived mediator of angiogenesis. Science 1992; 258:1798-1801.
127. Swensson O, Shubert C, Christophers E. Inflammatory properties of neutrophil-activating protein-1/interleukin-8(NAP-1/IL-8) in human skin: a light-and electromicroscopic study. J Invest Dermatol 1991; 96:682-689.
128. Colditz I, Zwahlen R, Dewald B et al. In vivo inflammatory activity of neutrophil-activating factor, a novel chemotactic peptide derived from human monocytes. Am J Pathol 1989; 134:755-760.
129. Vieira P, de Waal Malefyt R, Dang MN et al. Isolation and expression of human cytokine synthesis inhibitor factor (CSIF/IL-10) cDNA clones: homology to Epstein-Barr virus open reading frame BCRFI. Proc Natl Acad Sci USA 1991; 88:1172-1176.
130. Mosmann TR. Properties and functions of Interleukin-10. Adv Immunol 1994; 56:1-26.
131. Fiorentino DF, Bond MW, Mosmann TR. Two types of mouse helper T cell. IV. Th2 clones secrete a factor that inhibits cytokine production by Th1 clones. J Exp Med 1989; 170:2081-2095.
132. D'Andrea A, Aste Amezaga M, Valiante NM et al. Interleukin-10 inhibits human lymphocyte IFN-γ production by suppressing natural killer cell stimulatory factor/Interleukin-12 synthesis in accessory cells. J Exp Med 1993; 178:1041-1048.
133. Moore KW, O'Garra A, de Waal Malefyt R et al. Interleukin 10. Ann Rev Immunol 1993; 11:165-190.
134. de Waal Malefyt R, Abrams J, Bennett B et al. Interleukin 10(10) inhibits cytokine synthesis by human monocytes: an autoregulatory role of IL-10 produced by monocytes. J Exp Med 1991; 174:1209-1220.
135. Bogdan C, Vodovotz Y, Nathan C. Macrophage deactivation by Interleukin 10. J Exp Med 1991; 174:1549-1555.
136. Fiorentino DF, Zlotnik A, Mosmann TR et al. Interleukin 10 inhibits cytokine production by activated macrophages. J Immunol 1991; 147:3815-3822.
137. Fiorentino DF, Zlotnik A, Vieira P et al. IL-10 acts on the antigen-presenting cells to inhibit cytokine production by Th1 cells. J Immunol 1991; 146:3444-3451.
138. Hsieh CS, Heimberger AB, Gold JS et al. Differential regulation of T helper phenotype development by IL-4 and IL-10 in an alpha beta transgenic system. Proc Natl Acad Sci USA 1992; 89:6065-6069.
139. Trinchieri G. Interleukin-12: a proinflammatory cytokine with immunoregulatory functions that bridge innate resistance and antigen-specific adaptive immunity. Ann Rev Immunol 1995; 13:251-276.

140. Gerard C, Bruyns C, Marchant A et al. Interleukin 10 reduces the release of tumor necrosis factor and prevents lethality in experimental endotoxemia. J Exp Med 1993; 177:547-550.
141. Howard M, Muchamuel T, Andrade S et al. Interleukin 10 protects mice from lethal endotoxemia. J Exp Med 1993; 177:1205-1208.
142. Standiford TJ, Strieter RM, Lukacs NW et al. Neutralization of IL-10 increases lethality in endotoxemia. Cooperative effects of macrophage inflammatory protein-2 and tumor necrosis factor. J Immunol 1995; 155:2222-2229.
143. Kobayashi M, Fitz L, Ryan M et al. Indentification and purification of natural killer cell stimulatory factor (NKSF), a cytokine with multiple biological effects on human lymphocytes. J Exp Med 1989; 170:827-846.
144. D'Andrea A, Rengaraju M, Valiante NM et al. Production of natural killer cell stimulatory factor (interleukin 12) by peripheral blood mononuclear cells. J Exp Med 1992; 176:1387-1398.
145. Flesh IEA, Hess JH, Huang S et al. Early IL-12 production by macrophages in response to mycobacterial infection depends on IFNγ and TNFα. J Exp Med 1995; 181:1615-1621.
146. Kubin M, Chow JM, Trinchieri G. Differential regulation of interleukin-12, tumor necrosis factor-α, and IL-1β production in human myeloid leukemia cell lines and peripheral blood mononuclear cells. Blood 1993; 83:1847-1855.
147. Hayes MP, Wang J, Norcross MA. Regulation of Interleukin-12 expression in human monocytes: selective priming by interferon-γ of lipopolysaccharide-inducible p35 and p40 genes. Blood 1995; 86:646-650.
148. D'Andrea A, Ma X, Aste Amezaga M et al. Stimulatory and inhibitory effects of Interleukin-4 and IL-13 on production of cytokines by human peripheral blood mononuclear cells: priming for IL-12 and tumor necrosis factor-α production. J Exp Med 1995; 181:537-546.
149. Gately MK, Wolitzky AG, Quinn PM et al. Regulation of human cytolytic lymphocyte responses by interleukin-12. Cell Immunol 1992; 143:127-142.
150. Trinchieri G. Interleukin-12 and its role in the generation of Th1 cells. Immunol Today 1993; 14:335-338.
151. Gazzinelli RT, Hieny S, Wynn TA et al. Interleukin-12 is required for the T-lymphocyte independent induction of interferon-γ by an intracellular parasite and induces resistance in T-deficient hosts. Proc Natl Acad Sci USA 1993; 90:6115-6119.
152. Tripp CS, Wolf SF, Unanue ER. Interleukin-12 and tumor necrosis alpha are costimulators of interferon gamma production by natural killer cells in severe combined immunodeficiency mice with listeriosis, and interleukin 10 is a physiologic antagonist. Proc Natl Acad Sci USA 1993; 90:3725-3729.
153. Tripp CS, Gately MK, Hakimi J et al. Neutralization of IL-12 decreases resistance to *Listeria* in SCID and CB-17 mice. J Immunol 1994; 152:1883-1887.

154. Gazzinelli RT, Wysocka M, Hayashi S et al. Parasite induced IL-12 stimulates early IFNγ synthesis and resistance during acute infection with *Toxoplasma gondii*. J Immunol 1994: 153:2533-2543.
155. Vieira LQ, Hondowicz BD, Afonso LCC et al. Infection with *L.major* induces interleukin-12 production in vivo. Immunol Lett 1994; 40:157-161.
156. Romani L, Mencacci A, Tonnetti L et al. Interleukin-12 but not interferon-γ production correlates with induction of T helper type-1 phenotyppe in murine candidiasis. Eur J Immunol 1994; 22:909-913.
157. Zhang M, Gately MK, Wang E et al. Interleukin-12 at the site of disease in tuberculosis. J Clin Invest 1994; 93:1733-1739.
158. Chehimi J, Starr S, Frank I et al. Impaired IL-12 production in human immunodeficiency virus-infected patients. J Exp Med 1994; 179:1361-1366.
159. Derynck R, Balentien E, Han JH et al. Recombinant expression, biochemical characterization, and biological activities of the human MGSA/gro protein (published erratum appears in Biochemistry 1991 Jan 15:594). Biochemistry 1990; 29:10225-10233.
160. Schroder JM, Persoon NL, Christophers E. Lipopolysaccharide-stimulated human monocytes secrete, apart from neutrophil-activating peptide 1/interleukin-8, a second neutrophil-activating protein: NH2-terminal amino acid sequence identity with melanoma growth stimulatory activity. J Exp Med 1990; 171:1091-1100.
161. Moser B, Clark-Lewis I, Baggiolini M. Neutrophil-activating properties of the melanoma growth-stimulatory activity. J Exp Med 1990; 171:1797-1802.
162. Haskill S, Peace A, Morris J et al. Identification of three related human GRO genes encoding cytokine functions. Proc Natl Acad Sci USA 1990; 87:7732-7736.
163. Cao Y, Chen C, Weatherbee JA et al. Gro-β, a -C-X-C- chemokine, is an angiogenesis inhibitor that suppresses the growth of Lewis lung carcinoma in mice. J Exp Med 1995; 182:2069-2077.
164. Wolpe SD, Davatelis G, Sherry B et al. Macrophages secrete a novel heparin-binding protein with inflammatory and neutrophil chemokinetic properties. J Exp Med 1988; 167:570-581.
165. Fahey TJ III, Tracey KJ, Tekamp-Olson P et al. Macrophage inflammatory protein-1 modulates macrophage function. J Immunol 1992; 148:2764-2769.
166. Davatelis G, Wolpe SD, Sherry B et al. Macrophage inflammatory protein-1: a prostaglandin-independent endogenous pyrogen. Science 1989; 243:1066-1068.
167. Graham GJ, Wright EG, Hewick R et al. Identification and characterization of an inhibitor of haematopoietic stem cell proliferation. Nature 1990; 344:442-444.
168. Dunlop DJ, Wright EG, Lorimore S et al. Demonstration of stem cell inhibition and myeloprotective effects of SCI/rhMIP-1α in vivo. Blood 1992; 79:2221-2225.

169. Taub DD, Conlon K, Lloyd AR et al. Preferential migration of activated CD4+ and CD8+ T cells in response to MIP-1α and MIP-1β. Science 1993; 260:355-358.
170. Rollins BJ, Morrison ED, Stiles CD. Cloning and expression of JE, a gene inducible by platelet-derived growth factor and whose product has cytokine-like properties. Proc Natl Acad Sci USA 1988; 85:3738-3742.
171. Rollins BJ, Walz A, Baggiolini M. Recombinant human MCP-1/JE induces chemotaxis, calcium influx, and the respiratory burst in human monocytes. Blood 1991; 78:1112-1116.
172. Vaddi K, Newton RC. Regulation of monocyte integrin expression by β-family chemokines. J Immunol 1994; 153:4721-4732.
173. Rollins BJ. JE/MCP-1: an early-response gene encodes a monocytes-specific cytokine. Cancer Cells 1991; 3:517-524.
174. Jiang Y, Beller DI, Frendl G et al. Monocyte chemoattractant protein-1 regulates adhesion molecules expression and cytokine production in human monocytes. J Immunol 1992; 148:2423-2428.
175. Matsushima K, Larsen CG, DuBois GC et al. Purification and characterization of a novel chemotactic and activating factor produced by a myelomonocytic cell line. J Exp Med 1989; 169:1485-1490.
176. Colotta F, Borre' A, Wang JM et al. Expression of a monocyte chemotactic cytokine by human mononuclear phagocytes. J Immunol 1992; 148:760-765.
177. Van Damme J, Proost P, Lenaerts JP et al. Structural and functional identification of two human, tumor derived monocyte chemotactic proteins (MCP-2 and MCP-3) belonging to the chemokine family. J Exp Med 1992; 176:59-65.
178. Luster AD, Unkeless JC, Ravetch JV. Gamma interferon transcriptionally regulates an early response gene containing homology to platelet proteins. Nature 1985; 315:672-676.
179. Narumi S, Hamilton TA. Inducible expression of murine IP-10 mRNA varies with the state of macrophage inflammatory activity. J Immunol 1991; 146:3038-3044.
180. Taub DD, Lloyd AR, Conlon K et al. Recombinant human interferon-inducible protein 10 is a chemoattractant for human monocytes and T lymphocytes and promotes T cell adhesion to endothelial cells. J Exp Med 1993; 177:1809-1814.
181. Clark-Lewis I, Dewald B, Geiser T et al. Platelet factor 4 binds to Interleukin-8 receptors and activates neutrophils when its N terminus is modified with Glu-Leu-Arg. Proc Natl Acad Sci USA 1993; 90:3574-3577.
182. Kaplan G, Luster AD, Hancock G et al. The expression of a gamma interferon-induced protein (IP-10) in delayed immune responses in human skin. J Exp Med 1987; 166:1098-1108.
183. Gottlieb AB, Luster AD, Posnett DN et al. Detection of a gamma interferon-induced protein IP-10 in psoriatic plaques. J Exp Med 1988; 168:941-948.

184. Anderson DM, Lyman SD, Baird A et al. Molecular cloning of mast cell growth factor, a hematopoietin that is active in both membrane bound and soluble forms. Cell 1990; 63:235-243.
185. Martin FH, Suggs SV, Langley KE et al. Primary structure and functional expression of rat and human stem cell factor DNAs. Cell 1990; 63:203-211.
186. Galli SJ, Zsebo KM, Geissler EN. The Kit ligand, stem cell factor. Adv Immunol 1994; 55:1-96.
187. Matos ME, Schnier GS, Beecher MS et al. Expression of a functional c-kit receptor on a subset of natural killer cells. J Exp Med 1993; 178:1079-1084.
188. Flanagan JC, Chan DC, Leder P. Transmembrane form of the c-kit ligand growth factor is determined by alternative splicing and is missing in the Sld mutant. Cell 1991; 64:1025-1035.
189. Briddel RA, Hartley CA, Smith KA et al. Recombinant rat stem cell factor synergizes with recombinant human granulocyte colony stimulating factor in vivo in mice to mobilize peripheral blood progenitor. Blood 1993; 82:1720-1723.
190. Clark SC, Kamen R. The human hematopoietic colony stimulating factors. Science 1987; 236:1229-1237.
191. Nicola NA. Granulocyte-macrophage colony-stimulating factor. In: Nicola NA, ed. Guidebook to Cytokine and their Receptor. Oxford: Oxford Univ Press, Sambrook and Tooze Publication, 1994:171-173.
192. Gasson JC. Molecular physiology of granulocyte-macrophage colony-stimulating factor. Blood 1991; 77:1131-1145.
193. Demetri GD, Griffin JD. Granulocyte colony stimulating factor and its receptor. Blood 1991; 2791-2808.
194. Roth P, Stanley ER. The biology of CSF-1 and its receptor. Curr Topics Microbiol Immunol 1992; 181:141-167.

CHAPTER 3

MAIN CHARACTERISTICS OF CYTOKINE PRODUCTION BY HUMAN NEUTROPHILS

Although it was originally thought that monocytes and lymphocytes were the predominant sources of cytokines, it is now evident that the latter are generated by many other cell types as well. For instance, a large number of studies has established that mature polymorphonuclear leukocytes (PMN) also have the capacity to express the mRNA for, and subsequently secrete, several important inflammatory and immunoregulatory cytokines, either constitutively, or following appropriate stimulation. Although most of the studies addressing neutrophil cytokine release have been conducted in vitro, principally by the use of sensitive approaches such as immunohistochemistry (IH) and in situ hybridization (ISH), or by molecular biology techniques, in vivo studies have often confirmed the validity of the in vitro findings. This, obviously, points to new mechanisms by which PMN might significantly influence the multiple aspects of the inflammatory and immune responses.

In this chapter I shall present an overview on the general characteristics of cytokine production by neutrophils. The next chapters will focus on more specific aspects thereof, based on in vitro and in vivo studies.

IN VITRO PRODUCTION OF CYTOKINES BY HUMAN NEUTROPHILS: GENERAL FEATURES

Once isolated, neutrophils are usually suspended in culture medium containing antibiotics and variable concentrations [up to 10% of human or fetal bovine serum (FBS)], and then cultured at 37% in a 5% CO_2 humidified atmosphere. In our laboratory, we routinely

---

Cytokines Produced by Polymorphonuclear Neutrophils: Molecular and Biological Aspects, by Marco A. Cassatella. © 1996 R.G. Landes Company.

resuspend the PMN at no more than 5-10 x 10⁶/ml in RPMI 1640 medium containing 10% FBS (whose endotoxin content is <0.006 ng/ml). Other researchers dilute the PMN in serum-free medium or in classical buffers (phosphate buffered saline, Hanks's balanced salt solution, etc.) containing various concentrations of serum or bovine serum albumin, or none at all. I would not recommend the latter methods for long-term neutrophil cultures. In our conditions, cells can be cultured using plastic tissue culture flasks, or tissue culture plates, and then stimulated for up to 24 h. After the desired times of cultivation, the neutrophil suspension is harvested and then briefly centrifuged to separate the cell-free supernatants from the cell pellets. Cytokine synthesis or release can be measured in these two fractions by using several specific procedures, such as, for example, ELISA, radioimmunoassay (RIA), biological assays or immunoprecipitation after cell labeling. However, other techniques that have been used to detect cytokine expression are, as already mentioned, IH, ISH, Northern blot analysis and reverse-transcriptase polymerase chain reaction (RT-PCR).

As discussed in chapter 1, there is no longer any doubt that the release of cytokines constitutes an important aspect of neutrophil biology. The fact that neutrophils can synthesize, store, and release a wide array of cytokines should thus induce us to reconsider the general role of neutrophils in physiopathology.

Figure 3.1 depicts all the cytokines which, to date, have been shown to be released by PMN in vitro. Several groups have generated convincing evidence that PMN can release Interferon-α (IFNα),[1,2] Tumor Necrosis Factor-α (TNFα),[3-6] Interleukin-1α/β (IL-1α/β),[7-9] IL-1 receptor antagonist (IL-1ra),[10-12] IL-8,[13,14] IL-12,[15,16] Macrophage Inflammatory Protein-1α (MIP-1α) and MIP-1β,[17-19] Growth related gene product-α (GROα),[19-21] and Transforming Growth Factor-β (TGFβ).[22,23] In addition, mRNA expression or release by human PMN of Granulocyte Colony-Stimulating Factor (G-CSF),[24] Macrophage CSF (M-CSF),[24] IL-3,[25] GROβ,[26] CD30 ligand (CD30L),[27] Stem Cell Factor (SCF)/Kit ligand (KL)[28] and Interferon Inducible Protein-10 (IP-10)(our unpublished observations) have also been reported, albeit in single instances, and therefore require further confirmatory evidence. Finally, there are conflicting reports in the literature concerning the issue of whether IL-6, Granulocyte-Macrophage Colony-Stimulating Factor (GM-CSF), and Monocyte Chemotactic Protein-1 (MCP-1) expression can be induced in human neutrophils. Some reports have indicated that IL-6,[29] GM-CSF[25] and MCP-1[30] are expressed by human PMN, while studies from other groups did not confirm these observations.[6,13,31] In murine models however, three groups have reported that IL-6 can be synthesized by PMN.[16,32,33]

From a general point of view, the ability of PMN to produce such a variety of cytokines, suggests that these cells can significantly influence the direction and evolution not only of the inflammatory response,

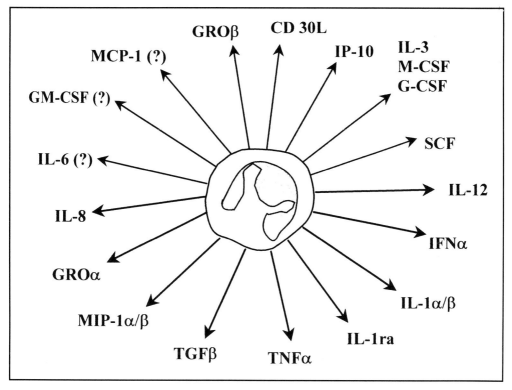

Fig. 3.1. Cytokines produced by human neutrophils in vitro. The question marks in brackets indicate the fact that no general consensus exists on the ability of human neutrophils to produce IL-6, MCP-1, or GM-CSF. For GROβ, CD30L and SCF, only mRNA expression has been reported to date.

but also of other processes, such as the immune response, antiviral defense, hematopoiesis, wound healing, and so forth.

STIMULI THAT INDUCE CYTOKINE PRODUCTION BY NEUTROPHILS

The production of individual cytokines by neutrophils is influenced to a great extent by the stimulus used. While the most widely used stimuli are bacterial lipopolysaccharide (LPS) and cytokines themselves, inducers of neutrophil cytokine generation can belong to many other categories of molecules or infectious noxae. Table 3.1 gives a partial list of agents that have been shown to stimulate PMN to produce cytokines. It is clear that other than LPS and cytokines, chemotactic factors [formyl-methionyl-leucyl-phenylalanine, fMLP), C5a, Leukotriene B_4 (LTB_4)], neuropeptides (substance P), phagocytosable particles, microorganisms such as fungi, viruses, and bacteria can also induce the release of cytokines by PMN. Although the current knowledge concerning the full ability of each stimulus to induce cytokine generation

Table 3.1. Ability of different stimuli to induce cytokine release by human neutrophils

	\multicolumn{11}{c}{Stimuli}										
	LPS	TNFα	IL-1	GM-CSF	IL-4	IL-13	fMLP	Substance P	Y-IgG	PMA	EBV
TNFα	+	+*		+*	-	-	-		+	+*	+
IL-1α/β	+	+	+	+	-	-	-		+	+	+
IL-1ra	+	+		+	+	+	+/-		+		
IL-8	+	+	+	+	-	-	+	+	+	+	
IL-12	+	-		+			-		-	-	
GROα	+	+		+			+/-		+		
MIP-1α	+	+		-	-	-	-*		+		

LPS: lipopolysaccharide; fMLP: formyl-Methionyl-Leucyl-Phenylalanine; Y-IgG: Saccharomyces cerevisiae opsonized with IgG; PMA: phorbol 12-myristate 13-acetate; EBV: Epstein Barr virus. *: mRNA expression only is rather limited, it is evident from Table 3.1 that the pattern of cytokines released by PMN is, in most cases, apparently stimulus-specific. For instance, IL-4 and IL-13 induce only IL-1ra, while substance P clearly induces only IL-8. On the other hand, phagocytosis of *S.cerevisiae* opsonized with IgG (Y-IgG) induces the release of many cytokines, but not that of IL-12, in contrast to LPS. In view of the fact that neutrophils represent the first line of defense against most invading agents and that their responses are stimulus-specific, a key role may be played by these cells in determining (or at least influencing) the evolution of the subsequent host responses.

Another important point is that the magnitude and kinetics of cytokine release vary substantially depending upon the stimulus used. For instance, Figure 3.2 shows an experiment in which IL-8 and IL-1ra were measured in the cell-free supernatants of cultured neutrophils stimulated with optimal concentrations of LPS, Y-IgG and fMLP. Under these conditions, IL-8[14] and IL-1ra,[11] together with MIP-1α/β[17,18] are clearly produced in larger amounts (in the order of nanograms/ml) compared with other cytokines (released in the order of picograms/ml). From figure 3.2 it is also evident that Y-IgG represents the most potent stimulus for IL-8 release, whereas the most powerful agonist for IL-1ra secretion is LPS. Furthermore, the kinetics of fMLP-stimulated IL-8 production strikingly differ from those of LPS or Y-IgG. That the fMLP-induced IL-8 release is maximal after 2-3 h and then declines to basal levels, is due to the fact that fMLP does not trigger the production of IL-1β or TNFα, contrary to LPS or Y-IgG. The ability of endogenous IL-1β and TNFα to stimulate the production of IL-8 in an autocrine manner (and thereby amplify total IL-8 production) will be described in greater detail below, as well as in chapter 5.

The differences observed among the respective actions of the various stimuli used in the experiment of Figure 3.2 also rule out any possible contamination of fMLP or Y-IgG with trace levels of endotoxin. Although this consideration might appears superfluous at first glance, it brings up an extremely important point. Endotoxin is, as mentioned above, a very potent stimulus of cytokine production in PMN; for instance, concentrations as low as 10 pg/ml LPS can induce IL-8 production (our unpublished observations). Unfortunately, LPS can easily contaminate solutions, reagents, labware, etc.,[34] unless stringent precautions are taken to prevent it. So, these facts must be kept always in mind. Using endotoxin-free serum and culture media, verifying the levels of LPS in each solutions by the *Limulus amebocyte lysate* (LAL) assay, and investigating whether polymyxin B sulfate inhibits the effects of the stimuli under investigation,[35] represent some of the common measures taken to keep LPS contamination under control. The latter method, if performed correctly, is particularly indicative of an eventual effect of contaminating LPS, since polymyxin B sulfate is highly effective in preventing the action of LPS.[35,36] However, it should

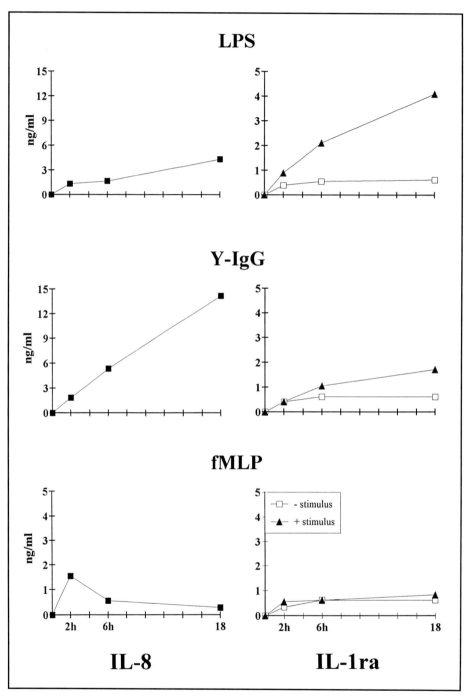

Fig. 3.2. Kinetics of the extracellular production of IL-8 and IL-1ra by human neutrophils stimulated with LPS (1 μg/ml), S.cerevisiae opsonized with IgG (Y-IgG) at a particle/cell ratio of 2/1, or fMLP (10 nM).

be emphasized that polymyxin B sulfate can also inhibit protein kinase C,[37] even in the same neutrophils.[38] An alternative to polymyxin B sulfate is the use of lipid A analogs.[39] As a rule, one should carefully investigate for the possibility of LPS contamination every time that a stimulus acts in similar manner to LPS. This would help in avoiding artifactual effects of some stimuli, which are sometimes reported in the literature. Conversely, should the solutions used to purify the neutrophils not be adequately controlled, they could easily pre-activate the cells[40,41] during isolation, and as a consequence, render them desensitized to a subsequent stimulation, for instance, with LPS. Endotoxin tolerance can be induced in a variety of primary cells or cell lines, including monocytes/macrophages,[42] and neutrophils in vivo.[43]

CYTOKINE GENE EXPRESSION

The induction of a cytokine production in PMN is usually preceded by an increased accumulation of the related mRNA transcripts. This can be detected by performing classical Northern blot analyses, but also by ISH or RT-PCR. Northern blotting has the advantage of making differences in cytokine mRNA expression readily visible, and (coupled to densitometric analysis) can be semi-quantitative. Although quantitative RT-PCR is also possible, I would recommend using it only after a previous demonstration that the results obtained are equivalent to those revealed by the Northern blot under identical conditions. This is because RT-PCR is so sensitive that one can easily amplify cytokine mRNA from a few (<0.5%) contaminating monocytes or lymphocytes.[44]

Figure 3.3 shows a typical Northern blot experiment, in which 10 μg of total RNA (purified from PMN treated for 4 h with different stimuli) were loaded on each gel lane. It is clear, for example, that, at the selected time-point, Y-IgG induces higher steady state levels of IL-8 mRNA than LPS or fMLP, but that LPS is by far the strongest inducer of mRNA accumulation for TNFα, IL-1β, IL-1ra and MIP-1α; LPS is also the only stimulus inducing IL-12p40 mRNA accumulation. From a general point of view, Northern blot analyses (and even more so RT-PCR) must be performed with adequate controls to avoid false positive results, which could be due to mononuclear cell contamination. I will examine this point in more detail in later sections (chapters 4 and 9), but let it be mentioned at this point that the extent of cytokine production by neutrophils is relatively low. Depending on the cytokine, neutrophils produce between 10- and 300-fold less cytokine than monocytes, on an individual cell basis (see ref. 45 for a review). It follows that it is absolutely mandatory to work with highly purified PMN populations. In our laboratory, a lot of time has been spent optimizing neutrophil isolation procedures (which routinely yields >99.5% pure PMN), as well as in producing an elaborate body of

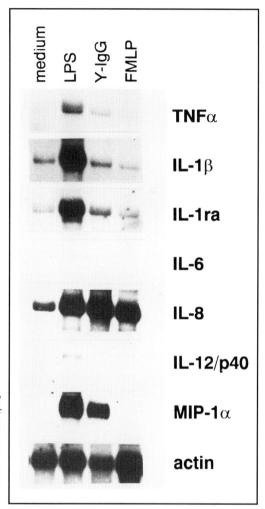

Fig. 3.3. Effect of LPS, Y-IgG and fMLP on the steady state levels of mRNA encoding various cytokines in human neutrophils. PMN were cultured for 4 h, and then subjected to Northern blot analysis. Actin mRNA expression is used to verify equal RNA loading.

evidence on the differential ability of PMN and monocytes to produce cytokines. If one considers that monocytes and lymphocytes possess 10-20 times more RNA per cell than neutrophils, a monocyte contamination of only 1% can translate into a total RNA preparation from PMN containing 20-25% contaminating monocyte RNA. Based on reports from our group,[6,20] as well as from other laboratories,[44,46] the presence of contaminating monocyte mRNA can be assessed by using IL-6 cDNA as a probe. As shown in Figures 3.3 and 3.4, neutrophils, unlike monocytes, do not express IL-6 transcripts. Thus, the absence of detectable IL-6 mRNA by Northern blot represents, in my opinion, a good control of PMN purity, because the presence of IL-6 mRNA could be an indication of contaminating monocytes in the

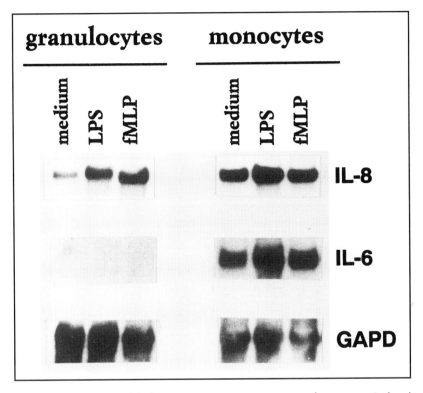

Fig. 3.4. Human neutrophils do not express IL-6 mRNA. PMN and monocytes isolated from the same donor were cultured with LPS or fMLP for 3 h, and then subjected to Northern blot analysis for IL-6 and IL-8 mRNA expression.

neutrophil populations. Other researchers have used a c-FMS cDNA probe for the same purpose,[24] but I consider IL-6 cDNA a much better indicator.

As will be discussed in chapter 9, there exist many other means to ascertain that one is working with highly pure neutrophils. For example, Figure 3.5 shows that following stimulation of monocytes and PMN purified from the same donor under identical experimental conditions, a subsequent assessment of the extracellular production of IL-8 reveals that the most potent inducer in neutrophils is Y-IgG, whereas in monocytes it is LPS.

The production of cytokines by neutrophils can be finely tuned, as demonstrated by the mechanisms underlying the synthesis and release of IL-12 and IP-10. In contrast to monocytes, in which a single stimulus induces the release of either cytokine, PMN produce IL-12 or IP-10 only if co-stimulated with at least two stimuli, for example LPS and IFNγ[15] (and our unpublished observations). The molecular basis for the requirement of two stimuli in order to induce IL-12 in

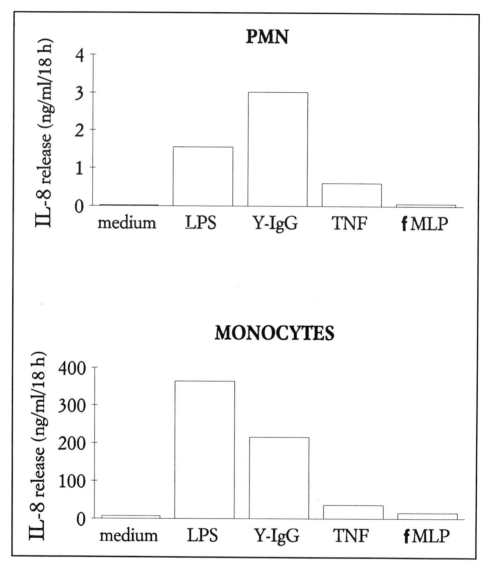

Fig. 3.5. Comparison of the ability of neutrophils and monocytes to produce IL-8. PMN (5 x 10⁶/ml) and monocytes (5 x 10⁵/ml) isolated from the same donor were cultured for 18 h with LPS, Y-IgG, fMLP or TNFα (5 ng/ml), before measuring IL-8 protein in their culture supernatants.

neutrophils was identified and is described in chapter 4;[15] for IP-10 however, further studies are required.

The use of Northern blotting and of related techniques can yield important insights into the molecular mechanisms regulating cytokine gene expression in PMN. The latter will be thoroughly described in chapter 6, but from a broad perspective, cytokine production in neutrophils can be regulated at transcriptional, post-transcriptional, translational and post-translational levels, as in other cell types.

CYTOKINE NETWORKS REGULATING NEUTROPHIL-DERIVED CYTOKINES

As will be discussed in detail in chapter 5, the production of cytokines by neutrophils can be modulated by immuno-regulatory cytokines such as Interferon-γ (IFNγ), IL-10, IL-4 and IL-13. Table 3.2, compiled mainly on the basis of studies from our laboratory, summarizes the effects of IL-10 and IFNγ on the production of various cytokines in human neutrophils. Depending on the stimulus used, the modulatory effects of IFNγ or IL-10 are different. For example, IFNγ potentiates the late release of IL-8 or GROα by LPS-stimulated PMN, but inhibits that induced by Y-IgG or fMLP. Similarly, IL-10 enhances the production of IL-1ra by LPS-stimulated PMN, but does not influence that induced by Y-IgG or fMLP. Finally, while IL-10 inhibits the production of GROα in LPS-stimulated PMN, it enhances that in response to TNFα.[20]

Interestingly, the in vitro production of IL-8 by human PMN in response to LPS seems to be regulated through a sort of cytokine network.[18,47,48] If one examines in detail the extracellular production of IL-8 in neutrophils stimulated with LPS (Fig. 3.6), it will be noted that PMN start to secrete IL-8 after 1 h. The production of IL-8 slowly increases up to 5-6 h, but, after this period, it dramatically increases, up to 18-20 h. The mechanisms underlying LPS-induced IL-8 release are depicted in Figure 3.7, which also raises the possibility that T helper type 1 (Th1) and Th2 lymphocytes[49,50] may influence the production of cytokines by PMN. Panel A shows that a direct effect of LPS-stimulation accounts for the initial release of IL-8, whereas the second wave is mediated mainly by the endogenous production of TNFα and IL-1β, which synergize (in an autocrine/paracrine manner) with LPS in inducing a late phase of IL-8 production. Several pieces of evidence support this network, including the effects of IL-10 and IFNγ, which can act in this circuit.

In PMN stimulated with LPS in the presence of IL-10 (Fig. 3.7, panel B), the initial production of IL-8 is unchanged, but the late IL-8 release is completely suppressed. This reflects the fact that IL-10 significantly inhibits the release of TNFα and IL-1β induced by LPS.[46,47] Indeed, when PMN are stimulated with LPS in the presence of neutralizing antibodies against TNFα and IL-1β, the late production of IL-8 is inhibited at levels comparable (but not identical) to those obtained when the cells are co-stimulated with IL-10 and LPS.[47] It cannot be excluded however that other endogenous neutrophil products may contribute to the late phase of IL-8 synthesis and secretion. Furthermore, IL-10 potentiates the release of the IL-1ra,[51] a cytokine known to counteract the binding of IL-1β to its receptors.[52]

Table 3.2. Regulation of extracellular cytokine production by IFNγ and IL-10 in neutrophils treated with different stimuli

	LPS				Y-IgG			fMLP	
	(−)	+ IFNγ	+ IL-10	(−)	+ IFNγ	+ IL-10	(−)	+ IFNγ	+ IL-10
TNFα	+	↑↑↑	↓↓	++	↑	→	−	−	−
IL-1β	++	↑↑	↓↓	+	↑	→	−	−	−
IL-1ra	++++	↑	↑↑	+	=	=	+/−	↑	=
IL-8	+++	↑↑	↓↓	+++++	→	→	+	→	=
IL-12p40	+	↑↑↑	↓↓	−	−	nd	−	nd	nd
GROα	+	↑	→	++	→	→	+/−	=	=
MIP-1α	++	↑	→	+++	nd	nd	−	nd	nd

LPS: lipopolysaccharide; Y-IgG: *Saccharomyces cerevisiae* opsonized with IgG; fMLP: formyl-Methionyl-Leucyl-Phenylalanine. Number of + reflects the extent of the response, whereas number of arrows reflects the extent of the up- or down-regulatory effect of IFNγ or IL-10. =: no effect; nd: not done.

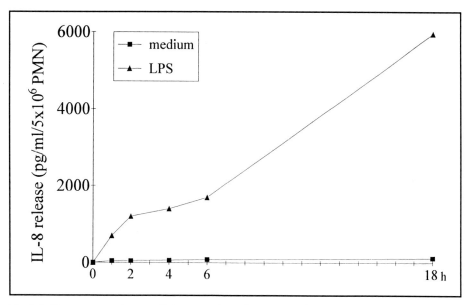

Fig. 3.6. Time course of extracellular IL-8 release by LPS-stimulated human neutrophils.

On the other hand, IFNγ strongly potentiates the production of TNFα and IL-1β, which results in a huge production of IL-8 at later times[53] (Fig. 3.7, panel C). Again, the second IL-8 wave is completely blocked by anti-TNFα and anti-IL-1β antibodies.[53] Finally, fMLP can induce the production of IL-8, but not that of TNFα or IL-1β[54]; consequently, there is no late wave of IL-8 release in response to fMLP.

Interestingly, the kinetics of GROα production in LPS-treated PMN, and the effects of IL-10 and IFNγ on this production,[20] are very similar to those observed in the case of IL-8. It is therefore tempting to speculate that GROα production in response to LPS might be regulated in a manner very similar to that of IL-8. Moreover, based on the results of Kunkel and coworkers,[18,48] a similar network probably also exists for MIP-1α and MIP-1β release in LPS-stimulated neutrophils. Those authors indeed reported that production of MIP-1α and MIP-1β not only features similar kinetics to those of IL-8, but that it is also affected by the action of neutralizing anti-TNFα antibodies, as well as by IL-10 and IFNγ, in a very similar manner to that described above for IL-8.[18,48]

CYTOKINE PRODUCTION BY NEUTROPHILS IN VIVO

Thus far, the production of cytokines by PMN has been focused exclusively on in vitro data. But what is known about the production of cytokines by neutrophils in vivo? Information on this topic is rapidly

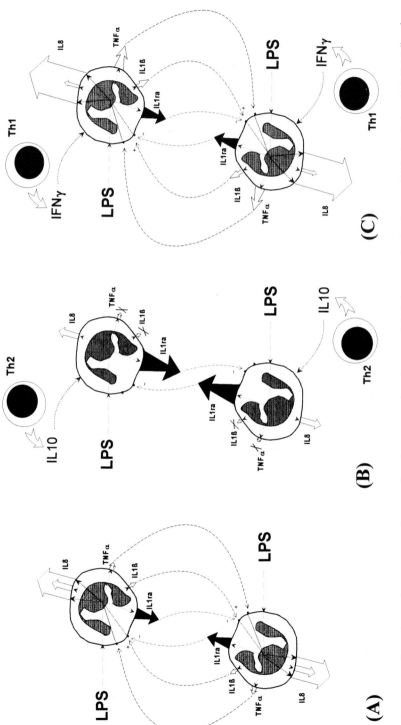

Fig. 3.7. Network regulating the in vitro production of IL-8 by LPS-stimulated neutrophils. (A) During the first hours of LPS stimulation, PMN initially release IL-8, and then TNFα and IL-1β. The latter cytokines have the potential to exert a feedback modulation (either on the same or on an adjacent neutrophil), resulting in an increased secretion of IL-8 at later (up to 20 h) incubation times. In this period, IL-1ra is also secreted in response to LPS. (B) IL-10 does not influence the initial release of IL-8 that is directly attributable to LPS, but suppresses the release of IL-1β and TNFα, and potentiates that of IL-1ra. In this manner, the late phase of IL-8 release is completely inhibited by IL-10. (C) IFNγ transiently suppresses the early release of IL-8, but potentiates the production of TNFα and IL-1β induced by LPS. As a consequence of this enhanced production of TNFα and IL-1β, the release of IL-8 is highly amplified at later times of LPS stimulation. Arrow size reflects the relative magnitude of cytokine release, whereas Th1 and Th2 refer to the lymphocyte subtypes.

growing, and many different experimental animal models have been developed. Figure 3.8 summarizes the cytokines which have been described to be produced by neutrophils in vivo, using animals injected intraperitoneally[32] or intravenously with LPS,[55] or instilled in the lung with LPS,[12,33,56] to induce specific acute inflammatory responses. Under these experimental conditions, it has been demonstrated, by northern analysis, PCR, IH or immunofluorescence of permeabilized cells, that TNFα, IL-1, IL-1ra, IL-6, IL-10, MIP-2 (functionally equivalent to human IL-8) and KC (functionally equivalent to GROα) can be expressed by neutrophils in vivo, and in some circumstances, even at levels higher than those found in mononuclear cells![12,33] This therefore not only confirms the validity of the observations made in vitro, but suggests that in vivo production of cytokines by neutrophils might have unsuspected pathogenetic consequences. The issue of in vivo production of cytokines by neutrophils is extensively treated in chapter 8.

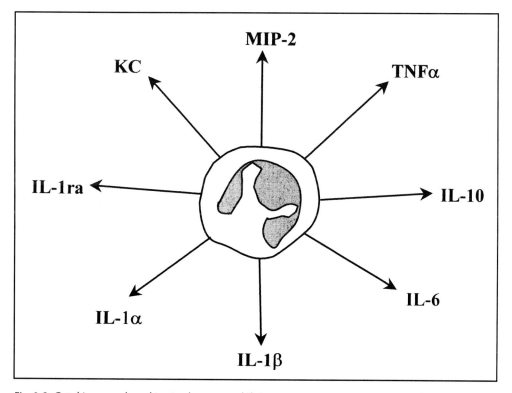

Fig. 3.8. Cytokines produced in vivo by neutrophils in response to LPS in experimental animal models.

REFERENCES

1. Shirafuji N, Matsuda S, Ogura H et al. Granulocyte-Colony-stimulating factor stimulates human mature neutrophilic granulocytes to produce interferon-α. Blood 1990; 75:17-19.
2. Brandt ER, Linnane AW, Devenish RJ. Expression of IFNα genes in subpopulations of peripheral blood cells. Br J Hematol 1994; 86:717-725.
3. Dubravec DB, Spriggs DR, Mannick JA et al. Circulating human peripheral blood granulocytes synthesize and secrete tumor necrosis factor alpha. Proc Natl Acad Sci USA 1990; 87:6758-6761.
4. Djeu JY, Serbousek D, Blanchard DK. Release of tumor necrosis factor by human polymorphonuclear leukocytes. Blood 1990; 76:1405-1409.
5. Mandi Y, Endresz V, Krenacs L et al. Tumor necrosis factor production by human granulocytes. Int Arch Allergy Appl Immunol 1991; 96:102-106.
6. Bazzoni F, Cassatella MA, Laudanna C et al. Phagocytosis of opsonized yeast induces TNFα mRNA accumulation and protein release by human polymorphonuclear leukocytes. J Leuk Biol 1991; 50:223-228.
7. Tiku K, Tiku ML, Skosey JL. Interleukin 1 production by human polymorphonuclear neutrophils. J Immunol 1986; 136:3677-3685.
8. Lindemann A, Riedel D, Oster W et al. Granulocyte-Macrophage Colony-Stimulating Factor induces Interleukin-1 production by humam polymorphonuclear neutrophils. J Immunol 1988; 140:837-839.
9. Lord PCW, Wilmoth LMG, Mizel SB et al. Expression of interleukin-1 α and β genes by human blood polymorphonuclear leukocytes. J Clin Invest 1991; 87:1312-1321.
10. Tiku K, Tiku ML, Liu S et al. Normal human neutrophils are a source of a specific interleukin 1 inhibitor. J Immunol 1986; 136:3686-3692.
11. McColl SR, Paquin R, Menard C et al. Human neutrophils produce high levels of the interleukin 1 receptor antagonist in response to Granulocyte/Macrophage Colony-stimulating factor and tumor necrosis factor. J Exp Med 1992; 176:593-598.
12. Ulich TR, Guo K, Yin S et al. Endotoxin-induced cytokine gene expression in vivo. IV. Expression of interleukin 1-α/β and interleukin 1 receptor antagonist mRNA during endotoxemia and during endotoxin-initiated local acute inflammatiom. Am J Pathol 1992; 141:61-68.
13. Strieter RM, Kasahara K, Allen R et al. Human neutrophils exhibit disparate chemotactic gene expression. Biochem Biophys Res Commun 1990; 173:725-730.
14. Bazzoni F, Cassatella MA, Rossi F et al. Phagocytosing neutrophils produce and release high amounts of the neutrophil activating peptide 1/Interleukin-8. J Exp Med 1993; 173:771-774.
15. Cassatella MA, Meda L, Gasperini S et al. Interleukin-12 production by human polymorphonuclear leukocytes. Eur J Immunol 1995; 25:1-5.
16. Romani L, Mencacci A, Cenci E et al. IL-12 as replacement therapy in neutropenic mice with fungal infection. 1996, submitted.
17. Kasama T, Strieter RM, Standiford TJ et al. Expression and regulation of human neutrophil-derived macrophage inflammatory protein 1-alpha. J Exp Med 1993; 178:63-72.

18. Kasama T, Strieter RM, Lukacs NW et al. Regulation of neutrophil-derived chemokine expression by IL-10. J Immunol 1994; 152:3559-3569.
19. Hachicha M, Naccache PH, McColl SR. Inflammatory mycrocrystal differentially regulate the secretion of macrophage Inflammatory protein 1 and Interleukin 8 by human neutrophils: a possible mechanism of neutrophil recruitment to sites of inflammation in synovitis. J Exp Med 1995; 182:2019-2025.
20. Gasperini S, Calzetti, F, Russo MP et al. Regulation of GROα production in human granulocytes. J Inflamm 1995; 45:143-151.
21. Koch AE, Kunkel SL, Shah MR et al. Growth-related gene product α. A chemotactic cytokine for neutrophils in rheumatoid arthritis. J Clin Invest 1995; 155:3660-3666.
22. Grotendorst GR, Smale G, Pencev D. Production of Transforming growth factor beta by human peripheral blood monocytes and neutrophils. J Cell Phys 1989; 140:396-402.
23. Fava RA, Olsen NJ, Postlethwaite AE et al. Transforming Growth factor β1 (TGFβ1) induced neutrophil recruitment to synovial tissues: implications for TGFβ-driven synovial inflammation and hyperplasia. J Exp Med 1993; 173:1121-1132.
24. Lindemann A, Riedel D, Oster W et al. Granulocyte-Macrophage Colony-Stimulating Factor induces cytokine secretion by humam polymorphonuclear leukocytes. J Clin Invest 1989; 83:1308-1312.
25. Kita H, Ohnishi T, Okubo Y et al. Granulocyte/Macrophage Colony-stimulating Factor and Interleukin 3 release from human peripheral blood eosinophils and neutrophils. J Exp Med 1991; 174:745-748.
26. Iida N, Grotendorst GR. Cloning and sequencing of a new *gro* transcript from activated human monocytes: expression in leukocytes and wound tissue. Mol Cell Biol 1990; 10:5596-5599.
27. Gruss HJ, DaSilva N, Hu ZB et al. Expression and regulation of CD30 ligand and CD30 in human leukemia-lymphoma cell lines. Leukemia 1994; 8:2083-2094.
28. Ramenghi U, Ruggieri L, Dianzani I et al. Human peripheral blood granulocytes and myeloid leukemic cell lines express both transcripts encoding for stem cell factor. Stem Cells 1994; 12:521-526.
29. Cicco NA, Lindemann A, Content J et al. Inducible production of Interleukin-6 by human neutrophils: role of Granulocyte-Macrophage Colony-Stimulating Factor and tumor necrosis factor alpha. Blood 1990; 75:2049-2052.
30. Burn TC, Petrovick MS, Hohaus S et al. Monocyte chemoattractant protein-1 gene is expressed in activated neutrophils and retinoic acid-induced human myeloid cell lines. Blood 1994; 84:2776-2783.
31. Contrino J, Krause PJ, Slover N et al. Elevated interleukin 1 expression in human neonatal neutrophils. Pediatr Res 1993; 34:249-252.
32. Terebuth PD, Otterness IG, Strieter RM et al. Biologic and immunohistochemical analysis of interleukin-6 expression in vivo. Constitutive and induced expression in murine polymorphonuclear and mononuclear phagocytes. Am J Pathol 1992; 140:649-657.

33. Xing Z, Jordana M, Kirpalani H et al. Cytokine expression by neutrophils and macrophages in vivo: endotoxin induces TNFα, MIP-2, IL-1β and IL-6 but not RANTES or TGFβ1 mRNA expression in acute lung inflammation. Am J Respir Cell Mol Biol 1994; 10:148-153: erratum Am J Respir Cell Mol Biol 1994; 10: following 346.
34. Haslett C, Guthrie LA, Kopaniak MM et al. Modulation of multiple neutrophil functions by preparative methods or trace concentrations of bacterial lypopolysaccharide. Am J Pathol 1985; 119:101-110.
35. Morrison DC, Jacobs DM. Binding of polymyxin B to the lipid A protein of bacterial lipopolysaccharide. Immunochemistry 1976; 133:813-820.
36. Serra MC, Calzetti F, Ceska M et al. Effect of substance P on superoxide anion and IL-8 production by human PMN. Immunology 1994; 82:63-69.
37. Mazzei GJ, Katoh N, Kuo JF. Polymyxin B is a more selective inhibitor for phospholipid-sensitive Ca^{2+}- dependent protein kinase than for calmodulin- sensitive Ca^{2+}-dependent protein kinase. Biochem Biophys Res Commun 1982: 109:1129-1133.
38. Naccache PH, Molski MM, Sha'afi RI. Polymyxin B inhibits phorbol 12-myristate 13-acetate, but not chemotactic factor, induced effects in rabbit neutrophils. FEBS Lett 1985; 193:227-230.
39. Christ WJ, Asano O, Robidoux A et al. E5531, a pure endotoxin antagonist of high potency. Science 1995; 268:80-83.
40. Kuijpers TW, Tool ATJ, van der Schoot CE et al. Membrane surface antigen expression on neutrophils: a reappraisal of the use of surface markers for neutrophil activation. Blood 1991; 78:1105-1111.
41. Venaille TJ, Misso NLA, Phillips MJ et al. Effects of different density gradient separation techniques on neutrophil function. Scand J Lab Invest 1994; 54:385-391.
42. Virca GD, KIM SY, Glaser KB et al. Lipopolysaccharide induces hyporesponsiveness to its own action in RAW 264.7 cells. J Biol Chem 1989; 264:21951-21956.
43. McCall CE, Grosso-Wilmoth LM, LaRue K et al. Tolerance to endotoxin-induced expression of the Interleukin-1β gene in blood neutrophils of humans with the sepsis syndrome. J Clin Invest 1993; 91:853-861.
44. Takeichi O, Saito I, Tsurumachi T et al. Human polymorphonuclear leukocytes derived from chronically inflamed tissue express inflammatory cytokines in vivo. Cell Immunol 1995; 156;296-309.
45. Cassatella MA. The production of cytokines by polymorphonuclear leukocytes. Immunol Today 1995; 16:21-26.
46. Wang P, Wu P, Anthes JC et al. Interleukin-10 inhibits Interleukin-8 production in human neutrophils. Blood 1994; 83:2678-2683.
47. Cassatella MA, Meda L, Bonora S et al. Interleukin 10 inhibits the release of proinflammatory cytokines from human polymorphonuclear leukocytes. Evidence for an autocrine role of TNFα and IL-1β in mediating the production of IL-8 triggered by lipopolysaccharide. J Exp Med 1993; 178:2207-2211.

48. Kasama T, Strieter RM, Lukacs NW et al. Interferon gamma modulates the expression of neutrophil-derived chemokines. J Invest Med 1995; 43:58-67.
49. Mosmann TR, Coffman RL. Th1 and Th2 cells: different patterns of lymphokine secretion lead to different functional properties. Ann Rev Immunol 1989; 7:145-173.
50. Romagnani S. Lymphokine production by human T cells in disease state. Ann Rev Immunol 1994; 12:227-257.
51. Cassatella MA, Meda L, Gasperini S et al. Interleukin 10 up-regulates IL-1 receptor antagonist production from lipolysaccharide-stimulated human polymorphonuclear leukocytes by delaying mRNA degradation. J Exp Med 1994; 179:1695-1699.
52. Dripps DJ, Brandhuber BJ, Thompson RC et al. Interleukin-1 (IL-1) receptor antagonist binds to the 80 kDa IL-1 receptor but does not initiate IL-1 signal transduction. J Biol Chem 1991; 266:10331-10336.
53. Meda L, Gasperini S, Ceska M et al. Modulation of proinflammatory cytokine release from human polymorphonuclear leukocytes by gamma interferon. Cell Immunol 1994; 57:448-461.
54. Cassatella MA, Bazzoni F, Ceska M et al. Interleukin 8 production by human polymorphonuclear leukocytes. The chemoattractant formyl-Methionyl-Leucyl-Phenylalanine induces the gene expression and release of interleukin 8 through a pertussis toxin sensitive pathway. J Immunol 1992; 148:3216-3220.
55. Williams JH, Patel K, Hatakeyama D et al. Activated pulmonary vascular neutrophils as early mediators of endotoxin-induced lung inflammation. Am J Resp Cell Mol Biol 1993; 8:134-144.
56. Xing Z, Kirpalani H, Torry D et al. Polymorphonuclear leukocytes as a significant source of tumor necrosis factor alpha in endotoxin-challenged lung tissue. Am J Pathol 1993; 143:1009-1015.

CHAPTER 4

PRODUCTION OF SPECIFIC CYTOKINES BY NEUTROPHILS IN VITRO

This chapter exclusively reviews the data regarding the production of cytokines by human neutrophils that, to date, have been accumulated from in vitro studies. Not included herein are eosinophil- and basophil-derived cytokine productions (see, for example, refs. 1 and 2 for reviews). The production of cytokines by polymorphonuclear leukocytes (PMN) in vivo will be discussed separately in chapter 8.

What follows is, hopefully, a complete description, cytokine by cytokine, of the stimuli or the experimental conditions observed to induce the production of cytokines by human neutrophils. Consideration of all of the biological effects of each neutrophil-derived cytokine would require a separate chapter for each molecule, so they are only briefly mentioned. The manner by which the release of each cytokine is influenced by modulatory agents, or is regulated at the molecular level, will be specific subjects, and treated separately in chapters 5 and 6, respectively.

INTERFERON-α (IFNα)

Table 4.1 summarizes the findings described to date regarding the stimuli able to induce the production of IFNα by neutrophils. The first exhaustive demonstration of the ability of neutrophils to express and release the IFNα protein was reported by Shirafuji et al.[3] Prior to that however, another study had shown that PMN-enriched fractions possessed very low levels of IFNα mRNA transcripts, as measured by S1-nuclease mapping, but contained no data at the protein level.[4] Shirafuji et al[3] attempted to determine whether neutrophils produce IFNα because earlier findings suggested that chronic myelogenous leukemia cells

---

Cytokines Produced by Polymorphonuclear Neutrophils: Molecular and Biological Aspects, by Marco A. Cassatella. © 1996 R.G. Landes Company.

Table 4.1. Effect of various agents on the production of IFNα or TGFβ by human neutrophils

Agent	Effect On IFNα (Ref.)		Effect On TGFβ (Ref.)	
	Undetected	Stimulating	Undetected	Stimulating
Medium				12
LPS	3		12*	
fMLP	3		12*	
G-CSF		3		
Sendai virus		6		
Immune complexes			12*	
PMA				13

LPS: lipopolysaccharide; fMLP: formyl-Methionyl-Leucyl-Phenylalanine; PMA: phorbol 12-myristate 13-acetate.
* : as compared to medium.

produce that factor.[5] By Northern blot analysis, they were able to demonstrate that G-CSF, in a time-dependent manner, induced the mRNA for IFNα type 1 in neutrophils, but not in mononuclear cells.[3] In addition, an IFNα-specific radioimmunoassay revealed that the levels of IFNα in the culture media of G-CSF-treated neutrophils rose, in a time-dependent manner, up to 100 U/ml per 10^7 cells after 12 h, whereas those of nonstimulated cells remained at about 20 U/ml.[3] Under the same experimental conditions, neither lipopolysaccharide (LPS) (10 ng/ml) nor formyl-methionyl-leucyl-phenylalanine (fMLP) (1 nM) effectively stimulated the expression of IFNα mRNA or protein in neutrophils.

Later studies confirmed and extended these early findings. By using the polymerase chain reaction (PCR), Brandt et al[6] were able to show that PMN express IFNα mRNA in a constitutive manner; IFNα-1, IFNα-2, and IFNα-4 species were those predominantly expressed. In addition, these various types of IFNα mRNA accumulated following infection with the Sendai virus.[6] A biological assay performed with cell-free culture supernatants revealed that the antiviral activity of PMN stimulated with Sendai virus for 6 h ranged from 63 to 1130 U/ml per 10^7 cells. These concentrations were similar to those detected in mononuclear cells, but much more abundant than those measured from purified T and B cells treated under similar experimental conditions.[6]

Taken together, these findings suggest that neutrophils can be significant sources of IFNα, and emphasize the potentially important role of PMN in host defense against viral infection. Although only limited attention has been paid to the role of neutrophils in viral infection, it is well known that they are important for protection against the influenza virus during the initial stage of infection in mice,[7] in diminishing the severity of vaccinia and herpes infections,[8] and in destroying human immunodeficiency virus (HIV).[9] Release of IFNα might be one of the mechanisms whereby neutrophils exert these functions. Considering that IFNα also increases cytokine-induced neutrophil-mediated anticandidal action of neutrophils,[10] the production of IFNα by PMN might stimulate these cells in an autocrine manner during candidiasis in HIV-immunocompromised patients. Furthermore, because IFNα was shown to inhibit neutrophil colony formation in vitro,[11] its release from PMN may be also viewed as a feedback regulatory mechanism for neutropoiesis.

TRANSFORMING GROWTH FACTOR-β (TGFβ)

Grotendorst and colleagues[12] were the first to report that cultured human neutrophils constitutively express TGFβ1 mRNA and secrete high levels of the protein. Secreted TGFβ was found to be structurally similar to that which is released by human platelets, and appeared to be in a fully active form, since acid treatment did not increase the amount of TGFβ activity,[12] as observed instead with other cell types (see chapter 2). It is possible that TGFβ released by PMN is already preactivated, or that it is activated outside the cell by the simultaneous discharge of proteolytic lysosomal enzymes. After activation of neutrophils or monocytes with LPS, fMLP or immune complexes for 24 h, no difference in the levels of TGFβ1 transcripts were detected, relative to untreated cells. We also observed in our laboratory that neutrophils constitutively express high levels of TGFβ1 mRNA, and that these levels are unaffected in LPS- or fMLP-activated cells (our unpublished observations). Remarkably, unstimulated PMN secreted approximately five times more TGFβ than an equal number of unstimulated monocytes, over a 24 h period in culture (approximately 50 ng/ml/10^6 PMN/24 h).[12] However, stimulation with LPS, fMLP or immune complexes resulted in a strong increase of TGFβ secretion by monocytes, as opposed to neutrophils.[12] Although the authors correctly hypothesized that the constitutive TGFβ secretion by neutrophils might have been due to their possible activation during the isolation procedure,[12] these results bring about additional considerations. First, the production of TGFβ in neutrophils and monocytes seems to be differentially regulated, and second, TGFβ production in monocytes might be controlled by post-transcriptional mechanisms.

That PMN may represent an important potential source of TGFβ in inflammatory infiltrates was confirmed by the studies of Fava et al.[13] Even though the authors did not consider the possible influence of contaminating monocytes and lymphocytes (up to 1%), they reported that a pre-existing TGFβ-bioactivity could be released from freshly isolated peripheral PMN, if incubated with 100 ng/ml phorbol-myristate-acetate

(PMA) at 37° for 30 min. The identity of this activity as TGFβ1 was immunologically confirmed, and 44% of the stored TGFβ1 was found to be released by PMN stimulated with PMA in the presence of cytochalasin B, a drug which facilitates degranulation.[14] In contrast with the findings of Grotendorst,[12] Fava et al[13] found a little detectable TGFβ1 activity released from unstimulated PMN, but this might be explained by the fact that neutrophils were only cultured for 30 min, as opposed to 24 h in the study of Grotendorst.[12] Table 4.1 summarizes the current knowledge on the neutrophil production of TGFβ. Worthy of note is that to date, no study has examined whether PMN specifically or differentially express TGFβ2 or TGFβ3 mRNA and proteins. This important aspect obviously requires further investigation, especially in view of the fact that TGFβ2 is the most abundant isoform detected in the synovial fluid (SF) of rheumatoid arthritis (RA) patients.[15] Since SF of RA patients contains large numbers of PMN,[16] it can be envisaged that TGFβ2 might be released from the SF PMN. Incidentally, the ability of neutrophils to produce TGFβ may also play an important role in situations such as wound repair, chronic immuno-driven inflammation and immune responses, or in the pathogenesis of fibrotic disease. In view of the recruiting ability, and of the many actions of TGFβ towards neutrophils or monocytes (succinctly listed in chapter 2) the release of TGFβ by PMN in vivo could additionally promote strong local effects.[17-19]

TUMOR NECROSIS FACTOR-α (TNFα)

The first molecular evidence of the ability of PMN to express TNFα mRNA was provided by Lindemann and colleagues.[20] They observed that PMN stimulated with GM-CSF for up to 24 h were induced to accumulate TNFα mRNA transcripts. However, neither the cell-free culture supernatants, nor the cell lysates demonstrated the presence of the TNFα protein, as determined by specific cytotoxic assays or using a radioimmunoassay (RIA) with detection limits of 0.1-0.5 ng/ml. On the basis of these results, the authors speculated that the expression of TNFα in PMN could be under a tight translational control. Since it was already known that in human peripheral blood monocytes, IFNγ was able to induce TNFα mRNA but not protein secretion unless an additional signal was provided,[21] such speculation was plausible. However, another report published almost in the same period,[22] described the involvement of TNFα in the granulocyte-mediated killing of two extremely TNFα-sensitive cell lines, namely WEHI sarcoma 164, and L929 cells. The specific cytotoxic activity of PMN against these two cell lines was markedly reduced by anti-TNFα neutralizing antibodies. Moreover, supernatants obtained from co-cultures of PMN and WEHI 164, but not from PMN alone, were effective in lysing ^{51}Cr-labeled targets, suggesting that a direct contact between effector granulocytes and the target cells was essential for the production of the lytic molecule(s).[22]

Direct and clear molecular evidence that PMN could release immunologically and biologically active TNFα was described in the studies performed by Dubravec et al,[23] and Djeu et al,[24] which were published almost simultaneously. The former group showed that in response to 5 µg/ml LPS, PMN exhibited increased levels of TNFα mRNA, with maximal accumulation after 2 h, and released approximately 160-190 pg/ml/5/ x 10^5 cells of TNFα into culture supernatants. They also showed that monocytes at a concentration 20 times higher than that potentially present in their purified PMN populations could not produce as much TNFα.[23] More recently, Haziot et al conferred to surface CD14 an important role in the secretion of TNFα by neutrophils in response to LPS.[25] They in fact demonstrated that stimulation of neutrophils under serum-free conditions with complexes of purified LPS-binding protein (LBP) and low concentrations of LPS (0.1 to 5 ng/ml), resulted (as with monocytes[26]), in a release of TNFα which was dependent upon CD14 expression, as anti-CD14 antibodies severely inhibited the response.[25]

Djeu et al[24] confirmed that LPS (1 to 1000 ng/ml) was a good inducer of TNFα from PMN. In addition, they reported that incubation of neutrophils with the opportunistic fungus, *C.albicans*, led to a substantial extracellular release of TNFα, peaking after 8 h of stimulation.[24] In addition, the release of TNFα in response to either *C.albicans* or LPS,[24] was found to be inhibitable by actinomycin D, an inhibitor of RNA synthesis (see also Fig. 6.1), as well as by emetine and cycloheximide (CHX), which block protein synthesis. Mandi et al[27] similarly found that human neutrophils can be stimulated by *C.albicans* to produce TNFα in vitro. Since PMN are crucial end-effector cells responsible for the elimination of *C.albicans,* with TNFα being a key activator of this PMN function,[28] the ability of PMN to produce TNFα suggests the possible existence of an autocrine feedback loop for self-activation, to muster host defense against such microbes.

Additional stimuli identified by Mandi et al,[27] as able to induce TNFα production by PMN included *E.coli, S.aureus, K.pneumonia*, LPS, but not PMA. Among these agents, heat-killed *S.aureus* induced the highest release of TNFα. We also showed that phagocytosis by PMN of *S.cerevisiae* opsonized with IgG (Y-IgG) induced an extracellular release of TNFα at levels much higher than those detected after treatment with endotoxin.[29] Phagocytosis of unopsonized *S.cerevisiae* (Y) also represents a very good stimulus for TNFα production by PMN, which, while less potent than Y-IgG, is nevertheless stronger than LPS.[30] Whether the engagement of Fcγ-receptors, FcγII and FcγRIII by Y-IgG (respective to Y) involves the generation of additional second messengers, which would explain the higher production of TNFα, remains to be investigated. Maximal TNFα recoveries in neutrophil supernatants in response to Y or Y-IgG,[29,30] as well as to LPS,[29] were detected after 5-6 h of stimulation, because at later time points the yields of TNFα were much lower. Similar kinetics of TNFα production were observed by other groups.[24,31] It is not yet clear why the yield of TNFα recovered from supernatants of stimulated neutrophils decays with time. Interestingly, phagocytosis carried out in the presence of protease inhibitors (such as α1-antitrypsin) increased the recovery of TNFα,[30] suggesting that the stability of TNFα in the culture medium is influenced by the release of proteolytic enzymes. This is in keeping with the ability of neutrophil elastase and cathepsin G to specifically degrade TNFα[32,33] but not IL-1α,[32] and with the fact that phagocytosis is a potent stimulus of neutrophil degranulation.[34] More recently,[35] the uptake of *Listeria monocytogenes* and *Yersinia enterocolitica* by PMN has also been shown to induce synthesis and secretion of large amounts of TNFα, suggesting that neutrophils recruited to sites of infection (liver, spleen, gastrointestinal tract) modulate the inflammatory and immune response in vivo during the onset of infection.

Another agonist recently identified as able to induce TNFα gene expression and secretion in PMN is IL-2.[36] The release of TNFα protein by IL-2-treated PMN was shown to be dose- and time-dependent, the increase in TNFα mRNA being maximal after 3 h of IL-2 exposure.[36] Furthermore, the induction of TNFα mRNA by IL-2 (1000 U/ml) was not inhibited by CHX, indicating that de novo protein synthesis was not necessary. In contrast, while the accumulation of TNFα mRNA by GM-CSF was inhibitable by CHX, that induced by IL-8 and heat-killed *C.albicans* was subject to superinduction by CHX, suggesting that, depending on the stimulus used, various induction pathways exist in PMN for the TNFα gene.[36] Interestingly, the release of TNFα protein by IL-2-treated PMN was inhibitable by specific monoclonal antibodies against the β chain of the IL-2 receptor (IL-2Rβ). In previous studies, the same group demonstrated that IL-2 could increase the antifungal activity of PMN, as well as the release of lactoferrin from these cells.[37] The expression of IL-2Rβ, but not of IL-2Rα, was identified on neutrophils, and shown to be involved in the IL-2 activation of the functions of PMN against *C.albicans*.[37] In more recent studies, Djeu and coworkers also reported that IL-2 prevents human neutrophil apoptosis in culture.[38] The cloning of the IL-2 receptor γ chain (IL-2Rγ) in Sugamura's laboratory demonstrated, however, that it is indispensable for the high- and intermediate-affinity IL-2 receptors, and that its cytoplasmic domain plays a critical role in IL-2-mediated intracellular signal transduction.[39,40] The latter group analyzed the expression of the IL-2Rα, -β and -γ chains in human peripheral blood cells, including granulocytes,[41] and found that most PMN express the γ chain, but not the α and β chains.[41] The lack of β chain expression in PMN was also observed by Philips et al.[42] The discrepancy with the data of Djeu et al[37] was attributed to nonspecific binding of antibodies, because Djeu et al[37] did not pretreat cells with serum or human IgG, which is essential for the specific staining of the β chain on granulocytes.[41] Expression of the γ chain in neutrophils has been further confirmed later by Djeu's group[43] and by Girard et al,[44] who used also northern blot analysis, as an experimental approach.

With regard to the IL-2/neutrophil relationship, I have to mention that we attempted several times to demonstrate either the presence of IL-2Rβ in freshly or cultured neutrophils, or a possible TNFα and IL-8 mRNA expression and production by human neutrophils stimulated with IL-2. Our experiments (still unpublished) gave negative results in all cases, even though we used three different commercial preparations of IL-2 (at doses ranging from 1 to 10000 U/ml), which were all shown to be effective in inducing lymphocyte proliferation and cytokine production from mononuclear cells isolated from the same donors used for neutrophil purification. We have no explanation for the discrepancy between our experiments and those of Djeu's group. Nevertheless, the recent observations of Girard et al[44] might provide a clue, insofar as they found that IL-2 alone did not modify the level of de novo RNA and protein synthesis in neutrophils,[44] similarly to what we observed. However, when the PMN were stimulated with IL-2 in combination with a fixed concentration of GM-CSF, but not of TNFα or fMLP, a dose-dependent effect of IL-2 in potentiating the induction of both RNA and protein synthesis by GM-CSF was observed. Although great donor-to-donor variability existed, these data suggest that IL-2 may actually function in neutrophils, but only if they are specifically "primed" by GM-CSF.

Table 4.2 reports the stimuli that induce TNFα expression in human neutrophils. As it can be noticed, some controversies concerning LPS exist, as in the case of IL-2. Takeichi et al,[45] for example, reported that neither LPS, nor Concanavalin A (Con A), nor zymosan were effective stimuli for TNFα production by neutrophils. However, they also noted that, in their experiments, TNFα mRNA was expressed. They attributed these conflicting results to the fact that their TNFα assay was not sensitive enough. Very recently, Terashima et al[46] also found that neutrophils do not produce TNFα after LPS-stimulation, even if they are coincubated with G-CSF. Despite these isolated reports, I believe that there is compelling evidence that PMN release TNFα in response to LPS, based also on more data generated by several other laboratories, including our own.[47-49] Finally, even sulfatides have been shown to induce TNFα (and IL-8) mRNA in neutrophils.[50] Since sulfatides have been established as ligands for L-selectins, it is tempting to speculate that ligation of L-selectin by PMN might activate their gene transcription. This could commit neutrophils to secrete cytokines and thus to better fulfill their proinflammatory role, once migrated into the extravascular space.

In any case, the ability of PMN to release TNFα in response to so many different stimuli, suggest that granulocytes may well exert host defense functions that go beyond the killing of invading microorganisms in septic infections. They could contribute not only to the

Table 4.2. Effect of various agents on the production of TNFα by human neutrophils

Agent	Effect (Ref.)	
	Undetected	Stimulating
LPS	46	23-25; 27; 29-31; 45*; 47-49
fMLP	77	
TNFα	77	20*
IL-1β	77	20*
IL-2	**	36
IL-8		36*
IL-10	120	
GM-CSF		20*; 36*
G-CSF	20*; 46	
IFNα	178	
IFNγ	47	
K. pneumoniae		27
S. aureus		27
Y. enterocolitica		35
L. monocytogenes		35
E. coli		27; 35
C. albicans		24; 27
Y-IgG		29; 30
Zymosan		45*
RSV	106	
PMA	27	
Con A		45*
Sulfatides		50*

Y-IgG: *Saccaromyces cerevisiae* opsonized with-IgG; RSV: Respiratory Syncytial virus;
Con A: Concanavalin A.
* : mRNA only.
** : our unpublished observations.

development and evolution of the inflammatory and immune responses, but also to tumor rejection.[51,52] As detailed in chapter 2, TNFα is a potent stimulus of PMN themselves, promoting adherence to endothelial cells and to particles, and leading to increased phagocytosis, respiratory burst activity and degranulation.[53-59] Therefore, the ability of PMN to produce TNFα may also represent a manner whereby neutrophils can activate themselves in an autocrine/paracrine manner.

INTERLEUKIN-1α AND -β (IL-1α AND -β)

IL-1 is a cytokine with multiple biological activities, which, accordingly, plays diverse roles in immunity and inflammation.[60] Although nowadays there is little doubt that neutrophils can be induced to produce small amounts of IL-1 under certain conditions in vitro (hundreds of picograms as a maximum), for a long time a considerable dispute has existed regarding the ability of PMN to synthesize and secrete IL-1. In the eighties, some investigators reported that neutrophils do not produce IL-1β.[61,62] The reasons for the lack of IL-1β detection may have been a poor sensitivity of the assays available at those times,[61,62] or inappropriate stimulatory conditions.[63] In addition, contaminating monocytes present in cell preparations have been proposed to account for the IL-1 production attributed to PMN.[61,62] However, observations made in 1984 by Goto et al,[64] on the capacity of murine and rabbit PMN harvested from peritoneal exudate cells induced by an intraperitoneal injection of casein, to produce an IL-1-like factor in response to kaolin (they termed it lymphocyte proliferation factor), stimulated many subsequent studies on the ability of human neutrophils to synthesize and release IL-1.

In 1986, Tiku et al[65] showed that human PMN (whose purity was more than 99%) produced an IL-1-like activity after stimulation with particulate and soluble agents, such as zymosan (1 mg/ml) and PMA (10 nM), respectively. This IL-1-like activity was detected in the supernatants of PMN only if the latter were stimulated for at least 4.5 h, and was also induced by calcium ionophore A23187 (500 nM) plus cytochalasin B (5 µg/ml), as well as by LPS (1 µg/ml), albeit at much lower levels.[65] Interestingly, while no IL-1-like activity was present in a preformed state or stored in PMN granules,[65] an indirect indication that cell lysates of resting PMN contained an endogenous inhibitor of IL-1 was provided.[65]

In later studies, human neutrophils stimulated with GM-CSF were observed to release IL-1. This was ascertained by Lindemann et al[66] who, by Northern blot analysis, detected the induction of mRNA specific for both IL-1α and IL-1β. Using standard biological assays, the same group also reported an IL-1 release that was dose- and time-dependent. These findings were strongly supported by the work of Dularay et al.[67] By using a sensitive IL-1 ELISA-SPOT assay able to enumerate and visualize the cells secreting IL-1 within a given population, these investigators confirmed that both IL-1α and IL-1β are produced by PMN in response to GM-CSF, but did not find the calcium ionophore A23187 (0.1-1 µM), cytochalasin B, or LPS (10-200 ng/ml) to be effective stimuli. Although PMN produced both IL-1α and IL-1β in response to GM-CSF, the number of cells producing IL-1α was greater and more strongly stained than that secreting IL-1β.[67] The ability of human neutrophils to synthesize and release IL-1 was further shown by Goh et al.[68] By either cytoplasmic RNA dot blot hybridization, or in situ hybridization analysis, these authors were able to detect transcripts for both IL-1α and IL-1β. Moreover, taking advantage of specific neutralizing antibodies to IL-1α and IL-1β, they also reported that in response to LPS (30 µg/ml), human PMN produce both forms of IL-1 over a 72 h culture period, and that IL-1β was produced and released before IL-1α.[68] Similar patterns of IL-1α and IL-1β production by neutrophils were very recently described by Takeichi et al.[45] The latter group showed that in response to zymosan (10-100 μg/m), LPS (50-100 μg/m) or Con A (10-25 μg/ml), the peak of IL-1β secretion by human PMN occurred at early periods (12-24 h), whereas IL-1α was secreted later (24-48 h). Neutrophils stimulated with LPS or Con A also secreted lower levels of IL-1α and IL-1β than those induced by zymosan.[45]

Lord at al[69] studied the expression of IL-1α and IL-1β in human PMN, in parallel with close monitoring of the effects of contaminating peripheral blood mononuclear cells (PBMC). They not only provided evidence that PMN transcribe and translate the IL-1α and IL-1β genes after stimulation with LPS (10 μg/ml) or IL-1α (100 U/ml), but also clearly established that IL-1α and IL-1β synthesis attributed to PMN could not be accounted for by the low level of contaminating PBMC.[69] Although virtually all PMN showed diffuse immunochemical staining with both IL-1α- and IL-1β-specific antisera, synthesis of IL-1β exceeded that of IL-1α, and little or no IL-1α was released by PMN over an 18 to 24 h time course.[69] The differences between the studies of Goh et al[68] and Lord et al,[69] on the one hand, and that of Dularay et al[67] on the other, may reside in the different doses of LPS used to trigger neutrophils, or in the different times of stimulation. Another very interesting observation made by Lord et al[69] was that although increases in IL-1 mRNA after stimulation of PMN and PBMC with LPS were similar, PMN were less efficient than PBMC in translating IL-1 mRNA. Since IL-1α and IL-1β mRNAs from either PMN or PBMC were translated with equal efficiency in rabbit reticulocyte lysates, Lord et al[69] speculated that synthesis of IL-1 in PMN is subject to some form of translational control. The fact that earlier on, Jack and Fearon[70] found PMN much less efficient than monocytes in synthesizing other proteins, suggests that neutrophils might have a generalized deficiency in translating their mRNA, probably due to their very low number of ribosomes.[71] In a subsequent paper,[72] McCall's group reported that blood neutrophils of patients with the sepsis syndrome (sepsis-PMN) are consistently tolerant to endotoxin-induced expression of the IL-1β gene, as defined by combined reductions in endotoxin-stimulated levels of IL-1β mRNA and decreased synthesis of the immunoreactive IL-1β protein. This down-regulation of the IL-1β gene in sepsis-PMN occurs concomitantly with an upregulation in the constitutive expression of the type II IL-1 receptor (IL-1RII),[72,73] and involves specific signal transduction pathways triggered by endotoxin, since sepsis-PMN respond normally to S.aureus induction of IL-1β synthesis. In addition, the down-regulation of the IL-1β gene in sepsis-PMN was not limited to infection by Gram-negative bacteria, but was also seen when the sepsis syndrome was apparently induced by

Gram-positive bacteria, *Rickettsia*, *Candida* species, or Staphylococcal exotoxins. The authors did not identify the precise mechanisms responsible for the tolerance of sepsis-PMN to endotoxin, but excluded that it was due to the loss of the CD14 surface protein (a receptor required for endotoxin gene induction in PMN),[25] or that it was the result of a global reduction in the functional responses of PMN.[72] The physiological significance of the tolerance to endotoxin for IL-1β gene expression may represent an attempt by the host to protect itself from the deleterious effects of disseminated inflammation.

Contributing more heat to the dispute of whether neutrophils produce IL-1 following endotoxin treatment were the findings of Hsi and Remick,[74] in spite of many other demonstrations by Wang et al,[48] Palma et al,[31] Malyak et al,[75] and my group.[30,47] In an ex vivo model of localized cytokine production that closely mimics in vivo conditions, that is, stimulation of human whole blood with LPS (1 μg/ml), Hsi and Remick found that PMN positivity for IL-1β by immunohistochemistry was very low and did not increase over time. In the same model, monocytes displayed a marked IL-1β signal which increased already by 4 h of LPS stimulation.[74]

Another group actively involved in studying the regulation of IL-1 synthesis by human PMN was the one led by D.L. Kreutzer. In 1990, he demonstrated that IL-1α, IL-1β and TNFα induced IL-1β mRNA accumulation in a dose-dependent manner.[76] Expression of IL-1β mRNA peaked at 1 h and returned to baseline levels by 2 h, but when PMN were treated with IL-1β and TNFα in combination, IL-1β mRNA levels remained elevated for up to 3 h.[76] The relative levels of the cell-associated antigenic IL-1β induced by TNFα and/or IL-1β strongly correlated with the induction of IL-1β mRNA, but no IL-1β antigen was detected in the supernatants of PMN incubated with or without cytokines for less than 4 h, as measured by a specific ELISA.[76] We also found that in response to TNFα, neutrophils produce and release IL-1β into the supernatant, but not before 6 h following stimulation.[47] In contrast with the findings of Marucha et al[76] and our own[47] for TNFα, and of Lindemann et al[66] and Dularay et al[67] for GM-CSF, Malyak et al[75] reported that treatment of neutrophils with TNFα (25 ng/ml) or GM-CSF (50 ng/ml), enhanced IL-1β mRNA levels transiently, but that did not result in an increase in IL-1β protein levels. Under identical experimental conditions, LPS was effective in inducing both IL-1β mRNA and protein release.[75] No comments or explanations for these discrepancies were provided by the authors,[75] but their results are intriguing, especially considering that they incubated PMN at 5×10^6/ml and used a very sensitive IL-1β ELISA (20 pg/ml). Kreutzer's group also reported that neonatal PMN can express antigenic IL-1β when stimulated by TNFα (20-1000 U/ml) and LPS (1 μg/ml), and that this expression is consistently higher than that of adult PMN.[77] In contrast, fMLP (100 nM) was found unable to induce

IL-1β in either neonatal and adult neutrophils. Whether the increased ability of neonatal PMN to majorly express IL-1β reflects a disfunction or an appropriate heightened response to the blunted immune response of the neonate, remains to be determined. Furthermore, in their last paper, Marucha and colleagues[78] reported that the increase of IL-1β mRNA expression induced by TNFα was not inhibited by difluoromethylornithine (DFMO), a selective inhibitor of ornithine decarboxylase.[78] On the contrary, DFMO inhibited the TNFα-induced rapid increases in putrescine and spermine content in PMN, and blunted the enhancement of superoxide generation and secondary granule release associated with priming by TNFα. Based on their results, the authors concluded that polyamine biosynthesis plays an important role in priming by TNFα, but is not involved in all the PMN responses to this cytokine, including IL-1β expression.[78]

Another condition inducing the mRNA expression and production of IL-1β by neutrophils is phagocytosis of yeast particles,[30] which provokes also the release of TNFα and IL-8. While kinetics of IL-1β, TNFα and IL-8 secretions in response to opsonized (Y-IgG) and unopsonized yeasts (Y) were similar, the absolute amounts of IL-1β detected in the supernatants of PMN stimulated with yeasts alone, interestingly, were usually higher than those obtained with Y-IgG-stimulated PMN, whereas the amounts of TNFα and IL-8 were lower.[30]

Included in the list of stimuli reported to determine the production of IL-1β from PMN, schematized in Table 4.3, are *Listeria monocytogenes* and *Yersinia enterocolitica*,[35] a mannoprotein constitutent (MP-F2) from *C.albicans*,[79] Epstein-Barr virus,[80] monosodium urate (MSU) and calcium pyrophosphate dihydrate (CPPD) crystals,[81] but not colchicine.[82] Finally, an apparent but artifactual increase in the amount of IL-1β released from activated neutrophils has been reported: this increase was the result of a technical problem, which interfered with the detection of IL-1β.[83] The observations that in so many situations neutrophils can produce IL-1α and IL-1β, could have physiopathological implications. IL-1 is a immunomodulatory and proinflammatory cytokine that is involved in host defense mechanisms, as well as in various acute and chronic disorders.[60] Release of IL-1 in concert with TNFα may fulfill immunomodulatory functions which may be beneficial for the host, but, if dysregulated, may contribute to the pathogenesis of the diseases in which IL-1 is involved.[60]

INTERLEUKIN-1 RECEPTOR ANTAGONIST (IL-1RA)

The first report suggesting that neutrophils might secrete endogenous products featuring IL-1-inhibitory activity were those of Tiku et al.[84] While studying the neutrophil production of an IL-1-like factor,[65] they observed that the addition of PMN to monocytes cultured in the presence of zymosan, led to a decreased biological IL-1 activity

Table 4.3. Effect of various agents on the production of IL-1α/β by human neutrophils

Agent	Effect on IL-1α (ref.)		Effect on IL-1β (ref.)	
	Undetected	Stimulating	Undetected	Stimulating
LPS	67	45; 48; 65; 68; 69	67	30; 31; 45; 47; 48; 65; 68; 69; 75; 77
fMLP				77
TNFα		82		47; 75*–78; 82
IL-1α				69; 76
IL-1β				76*
IL-2, IL-3–IL-6			76	
IL-10			120; 179	
GM-CSF		66; 67; 82		66; 67; 75*; 82
G-CSF			20*	
IFNγ			47	
S. aureus			68	72
Y. enterocolitica				35
L. monocytogenes				35
E. coli				27
C. albicans				79
Y-IgG				30
Zymosan		45; 46		45; 65
EBV				80
Con A		45		45
PMA				65
A23187+cytochalasin B	67	65	67	65
Colchicine			82	
MSU/CPPD				81; 82

EBV : Epstein Barr virus; MSU: monosodium urate; CPPD: calcium pyrophosphate dihydrate
* : mRNA only

of the resulting supernatants, relative to those harvested from monocytes stimulated in the absence of PMN. A possible explanation was that PMN released an inhibitor of IL-1.[65] A better characterization of this PMN-derived IL-1 inhibitory activity revealed that: (a), it was constitutively present either in lysates of freshly isolated PMN, or in cell-free supernatants obtained from PMN stimulated or not for 18 h; (b), it could be generated in the absence of serum; and, (c), it was not produced as a result of the activity of neutrophil-proteases.[84] Furthermore, the same studies pointed out that PMN contained a greater amount of this IL-1 inhibitory activity, than IL-1.[84] The latter observations provided an explanation for the difficulties reported by various investigators in detecting the production of IL-1 by PMN, especially when measured in biological assays.

The molecular demonstration that the IL-1 inhibitory activity corresponded to the IL-1ra produced by monocytes, appeared shortly after the cloning of the IL-1ra.[85,86] McColl et al,[87] by using both antibodies that were specific for IL-1ra, and an IL-1ra encoding cDNA, were able to show that neutrophils constitutively produce IL-1ra. This was in accordance with the findings reported previously by Tiku et al[84] using different experimental approaches. McColl et al[87] also showed that following activation with TNFα and GM-CSF, neutrophils secreted increased amounts of IL-1ra protein, in time-dependent manner, over a 24 h period (approximately 0.5 ng/ml/10^6 cells). In addition, they convincingly excluded any influence of contaminating monocytes, which are potent producers of IL-1ra. They estimated that monocytes are able to produce approximately 20 times more IL-1ra than neutrophils. Therefore, a minimum of 5% contaminating monocytes (instead of the 0.2% found in their PMN populations) could have accounted for all of the IL-1ra they measured in PMN supernatants.[87] That all these in vitro observations are likely to occur in an inflammatory setting in vivo, was demonstrated by the observation that neutrophils isolated from the synovial fluid of patients with RA clearly responded to GM-CSF and TNFα in terms of IL-1ra synthesis.[87] Worthy of note, and based mainly on previous findings reported by the same group,[88] the authors concluded that IL-1ra constitutes one of the major de novo-synthesized product of activated neutrophils, again extending prior observations made by Tiku et al[84] five years earlier. It was calculated that in neutrophils, IL-1ra is produced in excess of IL-1 by a factor of at least 100; such a ratio fits perfectly with the reported amounts of IL-1ra needed to inhibit the proinflammatory effects of IL-1.[60] In a more recent study in which both IL-1 and IL-1ra were measured in neutrophils,[82] the same group confirmed that GM-CSF and TNFα induce the production of 300 and 200 times more IL-1ra than IL-1, respectively.

Very surprisingly, in the same paper published in 1992, McColl et al[87] found that none of a large array of classical neutrophil stimuli (including fMLP, G-CSF, IL-1β, IL-4, TGFβ, IFNγ, and even LPS) influenced the extracellular production of IL-1ra by PMN. All of them were used over a 24 h period, and at concentrations ranging from sub- to supra-optimal doses. Although the results regarding several of those agonists, and especially LPS, have not been later reproduced by other investigators (see below), the reasons for which McColl and co-workers were unable to identify LPS as effective stimulus for the induction of IL-1ra synthesis by neutrophils remain intriguing. It should be said that in another paper previously published by the same group,[89] they again described LPS (used at doses up to 10 μg/ml) as unable to trigger nuclear signaling events necessary to induce de novo RNA synthesis in PMN. That finding was even more surprising, because several investigators had already recognized LPS as an efficient agonist for the modulation of specific mRNA expression in PMN.[24,29,69] Possible factors that could have hampered the detection of any modification of neutrophil gene expression in response to LPS, are the conditions used to isolate and culture PMN, which could have inadvertently preactivated the cells, and, as a consequence, desensitized them. A further possibility to consider is that McColl and coworkers cultured neutrophils in serum-free culture medium.[87,89] We also found that LPS below 100 ng/ml does not induce IL-8 in serum-free medium.[90] It is well known by now that in order to stimulate phagocytes at low doses, LPS necessitates to bind some different serum proteins, which in turn bind to CD14[25] (see also ref. 91). However, it has also been demonstrated that high doses of LPS (>100 ng/ml) can bypass the requirement for serum proteins, and are able by themselves to stimulate efficiently the target cells.[90] Finally, it is also possible that the particular batch or strain of LPS used by McColl's group was not properly working.

Ulich et al,[92] in a study published almost simultaneously to that of McColl et al,[87] found PMN able to express increased levels of IL-1ra mRNA and protein after in vitro incubation with LPS (100 μg/ml). The extent of IL-1ra mRNA expression found in PMN was in magnitude similar to, or even greater than, that found in PBMC. Analogous observations were subsequently made by Re et al,[93] and, as also shown in Figure 9.1, by us. In support of the relevance of the observations made in vitro, Ulich and colleagues also demonstrated that after intratracheal injection of LPS in the rat, a great expression of IL-1ra and IL-1β mRNAs could be observed in PMN-rich bronchoalveolar lavage.[92]

That LPS represents an effective stimulus for the production of IL-1ra by PMN was further demonstrated by other groups. We found, for instance, that LPS is not only a very efficient stimulus for the production of IL-1ra in PMN, but also that it is much more potent than Y-IgG (Fig. 4.1).[94] LPS increases the accumulation of IL-1ra into cell-free supernatants as early as after 2 h, and gives maximum yields at 20 h (Fig. 4.1).[94] In contrast, significant amounts of IL-1ra produced in response to Y-IgG were detected only 20 h after Y-IgG-phagocytosis, but at very low levels, and at least three to five times lower than those measured in supernatants of LPS-treated PMN (Fig. 4.1).[94] Consistent with a poor ability of Y-IgG to induce IL-1ra in neutrophils, Malyak et al[75] also reported that adherent IgG did not trigger the production of IL-1ra from PMN. It is noteworthy that the stimulation of human monocytes with LPS or IgG, these latter either in a soluble form,[95] or attached to plastic surfaces,[96] led instead to approximately the same amount of IL-1ra production. These results suggest that the ability of PMN and monocytes to secrete IL-1ra in

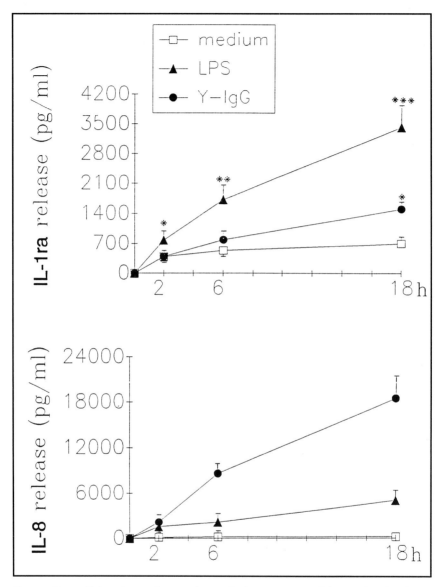

Fig. 4.1. Extracellular release of IL-8 and IL-1ra by human neutrophils stimulated with LPS or S.cerevisiae opsonized with IgG (Y-IgG).

response to specific agonists, acting presumably through identical receptors (the Fcγ-receptors), is regulated by distinct mechanisms. Whether this difference between monocytes and neutrophils is due to a greater ability of the monocyte Fcγ-receptors to generate intracellular signals for expression, translation and secretion of IL-1ra, remains to be investigated. In this respect, preliminary data indicate that adherent anti-FcγRIII (CD16) monoclonal antibodies, but not anti-FcγRI or anti-FcγRII monoclonal antibodies, induce the production of IL-1ra from PMN,[97] as measured in an IL-1 inhibitory assay.

Malyak et al[75] also confirmed the early findings of Tiku's group,[87] that freshly isolated PMN consistently contain a significant amount of cell-associated IL-1ra protein in the absence of detectable mRNA. The total levels of IL-1ra protein, as well as the abundance of IL-1ra mRNA, did not change when PMN were cultured in medium for up to 22 h, suggesting that their isolation procedures and the long-term culture in medium alone did not inadvertently lead to the stimulation of PMN. Moreover, they found that although IL-4 increased minimally the synthesis of IL-1ra by itself, it synergized with LPS in this action.[75] As examined by PCR, no intracellular forms of IL-1ra (icIL-1ra) mRNA were found in PMN, whether or not the cells were treated with LPS or GM-CSF.[75] Furthermore, through metabolic labeling experiments and carbohydrate digestion, it was shown for the first time that only the glycosylated form of IL-1ra is present in the culture supernatants of PMN.[75]

Another study showing that LPS concentrations as low as 10 ng/ml were sufficient to induce IL-1ra mRNA expression in PMN was that made by Re et al.[93] Furthermore, they reported that in PMN stimulated for 6 h, GM-CSF, G-CSF and IL-4 also increased the IL-1ra mRNA steady-state levels, while after a period of 18 h, IL-1ra was found to be secreted in response to LPS, GM-CSF and IL-4, but not in response to IL-1β, IFNγ, or chemotactic factors (fMLP, C5a and IL-8).[93] More recently, the same group provided strong evidence that IL-13[98] and TGFβ1[99] can also directly augment IL-1ra mRNA expression and protein production from human PMN. Both IL-13 and TGFβ1 prolonged IL-1ra mRNA half-life in neutrophils.[98,99] In addition, PCR analysis revealed that either IL-13[98] or TGFβ1[99] induce the mRNA expression of the intracellular form of IL-1ra (icIL-1ra) in PMN, which is usually in extremely low abundance or undetectable in myelomonocytic cells. The observation that TGFβ1 is able to trigger the production of IL-1ra, might explain one of the mechanism(s) whereby IL-1ra is constitutively produced by PMN.[84,87,94] As already mentioned, TGFβ is also constitutively expressed by PMN[12], and could therefore stimulate PMN to produce IL-1ra through an autocrine loop. Very recently, Muzio et al[100] cloned and characterized a new intracellular isoform of IL-1ra expressed in PMN and other cells, which was termed icIL-1ra type II (icIL-1raII). icIL-1raII was found to be potentially biologically active, and to differ from the previously known icIL-1ra (icIL-1raI) only by an additional stretch of 21 amino acids located within the NH$_2$-terminal portion of the molecule.[100]

The most recent study on the production of IL-1ra by PMN, performed by Marie et al,[101] confirmed that LPS, TNFα, IL-4, and TGFβ, could significantly enhance the spontaneous production of IL-1ra by human PMN, but in contrast to the findings of Re et al,[98] IL-13 was found to have a very weak capacity to induce IL-1ra production. In addition, only IL-4 (but not IL-10, TGFβ, or IL-13) enhanced the production of IL-1ra induced by LPS. However, when TNFα was used as a stimulus, both IL-4 and IL-10 synergistically amplified the induction of IL-1ra, while IL-13 and TGFβ had, again, no significant effect. IL-4 and IL-10 amplified both the release of IL-1ra and its cell-associated form.[101] Table 4.4 summarizes the list of the agents able to induce IL-1ra production in PMN. These also include the Epstein Barr virus,[80] but not MSU, CPPD or colchicine.[82]

Table 4.4. Effect of various agents on the production of IL-1ra by human neutrophils

Agent	Effect (ref.)	
	Undetected	Stimulating
Medium		84, 87, 94
LPS	87	75, 92-94, 101
fMLP	87, 93	
TNFα		75, 82, 87, 101
IL-1β	93, 87	
IL-3	87	
IL-4	87	75, 93, 101
IL-6	87	
IL-8	93	
IL-10	94, 179	
IL-13	101	98
GM-CSF		75, 82, 87, 93
G-CSF	87	93*
TGFβ	87	99, 101
IFNγ	87, 93	
Y-IgG		94
Zymosan		84
EBV		80
C5a	93*	
Adherent IgG	75	
anti-CD16 (FcγRIII)		97
anti-CD32 (FcγRII)	97	
anti-CD64 (FcγRI)	97	
MSU/CPDD	82	
Colchicine	82	

* : mRNA only.

As IL-1ra blocks the activities of IL-1 both in vitro and in vivo,[102] the production of relatively large amounts of IL-1ra by neutrophils could be of major biological significance since it could inhibit an established inflammatory response mediated by IL-1.

INTERLEUKIN-6 (IL-6)

As mentioned in chapter 3 and evidenced in Table 4.5, the question of whether human PMN produce IL-6 is still a matter of discussion. This controversy started in fact with the paper published by Cicco et al.[103] They reported that stimulation of PMN with GM-CSF (50 ng/ml), and to a lesser extent with TNFα (50 ng/ml), but not with G-CSF, M-CSF, IL-3, lymphotoxin, IFNγ, or fMLP, leads to a rapid accumulation of IL-6 mRNA, and subsequent secretion of the related protein. The latter was detected by a biological assay which was spe-

Table 4.5. Effect of various agents on the production of IL-6 by human neutrophils

Agent	Effect (ref.)	
	Undetected	Stimulating
Medium	45	31*, 35*, 105*
LPS	29, 45, 48, 104*	31, 103*
fMLP	103	
TNFα	**	103
Lymphotoxin	103	
IL-3	103	
GM-CSF		103, 105*
G-CSF	103	
M-CSF	103	
IFNγ	103	
Y. enterocolitica		35
L. monocytogenes		35
E. coli		35
C. albicans		79
Y-IgG	29	
Zymosan	45	
RSV		106
PMA	**	103*
Con A	45, 104*	
CHX	**	103*

CHX: cycloheximide.
* : mRNA only.
** : our unpublished observations.

cifically inhibited by anti-IL-6 polyclonal antibodies.[103] LPS (10 µg/ml), PMA (50 nM) and CHX (20 µg/ml) were also reported to induce IL-6 transcripts in PMN.[103] The authors, based on the cell quantities needed to obtain comparable hybridizing signals in northern blots, calculated that the specific capacity per cell was at least ten times less in PMN as compared with monocytes.[103] In contrast with the results of Cicco et al,[103] Kato and colleagues[104] previously showed that in a whole blood culture system used to identify the cells expressing IL-6 gene and protein after stimulation with LPS (1 µg/ml) and Con A, the IL-6-expressing cells were exclusively monocytes. This was demonstrated by a combination of techniques including immunofluorescence, immunocytochemical staining, and in situ hybridization. In accord with the data of Kato's group, we provided strong evidence for a complete lack of IL-6 gene expression in PMN cultured for up to 18 h in the presence of LPS (1 µg/ml) or Y-IgG.[29] An unequivocable support of our conclusions was the demonstration that in monocytes isolated from the same donors, IL-6 mRNA expression and secretion were strongly induced by both LPS and Y-IgG[29] (see also Figs. 3.4, and 9.1). On the basis of these findings, we consider the presence of IL-6 mRNA as a marker for monocyte contamination, and we screen every Northern blot made with RNA isolated from PMN for IL-6 expression.

After our study, other articles have appeared in the literature, showing that PMN express IL-6. Palma et al[31] for example, reported a time-dependent extracellular production of IL-6 in response to 100 ng/ml LPS, with a maximal release after 24 h (400 pg/ml/10^6 cells). This secretion was preceded by an augmentation of IL-6 mRNA by LPS, peaking after 1 h.[31] Remarkably, PMN also displayed a substantial expression of IL-6 transcripts under non stimulatory conditions.[31] Melani et al[105] also found a constitutive IL-6 expression in circulating human granulocytes immediately purified from fresh blood (98% neutrophils, 1.2% eosinophils, 0.8% mononucleated cells). For this investigation they used either reverse transcriptase-PCR (RT-PCR) (to detect low levels of RNA) or in situ hybridization (to confirm morphologically the cell type expressing the gene). Interestingly, the constitutive IL-6 expression was very weak if neutrophils were isolated from buffy coats instead of from fresh heparinized peripheral blood. Furthermore, IL-6 expression disappeared if PMN were left on ice or at 37°C in Dulbecco's modified Eagle's medium for 1 h, but it was still inducible by the addition of GM-CSF (50 ng/ml) for another hour.[105] Among several possible explanations, the authors[105] could not rule out whether the handling procedures, or the different anticoagulants used to purify neutrophils, specifically affected the constitutive or the stimulated expression of IL-6. In another paper in which the PMN from subjects at different stages of HIV infection were assayed, Cassone et al[79] reported that neutrophils produced significant amounts of IL-6 (340-430 pg/ml/18 h) and IL-1β after an overnight stimulation with MP-F2, a mannoprotein from *C.albicans*. However, there were no differences between HIV-negative and HIV-positive PMN. Finally, Arnold and colleagues published two papers on the ability of PMN to produce IL-6. In the first one,[35] they reported that after phagocytosis of *L.monocytogenes*, *Y.enterocolitica* and *E.coli*, PMN secreted large amounts of IL-6, IL-1β and TNFα. Furthermore, they also reported a constitutive IL-6 mRNA expression in resting neutrophils.[35] In the second one,[106] they showed that after exposure to infectious respiratory syncytial virus (RSV) particles, PMN secreted both IL-6 and IL-8, but not TNFα. Production of IL-6 occurred in a dose- and time-dependent manner, peaked at 3-5 h (1000 pg/ml/5 x 10^6 cells) and declined to undetectable levels at 24 h.[106] It should be noted that the same authors stated that their neutrophil preparations were less than 98% pure,[35,106] and this, in my opinion, could have significantly affected their results.

More recently, and supporting the data of Kato et al[104] as well as our own,[29] two new studies have emphasized that if production of IL-6 is detected in the cultures of human PMN, it is accounted for by contaminating monocytes. In one of them, Wang et. al[48] demonstrated that if neutrophils are prepared using standard procedures so that monocyte contamination is of 0.8% or more, they release significant amounts of IL-6 into the supernatants. Conversely, if extreme caution during PMN isolation is taken, for example in removing the Ficoll-Paque phase or in cleaning with a gauze the wall tube to reduce monocyte contamination below 0.7%, IL-6 release from PMN becomes undetectable. What is more, in experiments performed with cells isolated from the same donor, they also show, as we did,[29] that after LPS stimulation only monocytes (and not PMN) express IL-6 mRNA.[48]

In the second study, Takeichi et al[45] characterized cytokine secretion by peripheral blood PMN, monocytes and lymphocytes stimulated with zymosan, Con A and LPS. Both monocytes and lymphocytes produced high levels of IL-1α, IL-1β, IL-6 and TNFα, while PMN produced low levels of IL-1α, IL-1β, but not IL-6 or TNFα. In particular, the cells were also examined for cytokine gene expression by a very sensitive radioactive RT-PCR analysis. PMN stimulated with zymosan showed stronger expression of IL-1α, IL-1β, and TNFα mRNA than cells stimulated with LPS or Con A, but IL-6 mRNA was never detected under any condition. Furthermore, identical results were observed if PMN were isolated from the gingival crevicular fluid from patients affected by periodontitis.[45] In contrast, stimulated monocytes or lymphocytes expressed all cytokine transcripts at very high levels.[45] It was therefore concluded that PMN express only IL-1α, IL-1β and TNFα, but not IL-6 mRNA, and that the mRNA detected in this type of cells clearly derived from PMN and not from contaminating monocytes or lymphocytes.[45]

In summary, although it cannot be definitively and formally ruled out that certain stimuli might specifically induce IL-6 from human

PMN, the compelling evidence accumulated by now emphasizes the need to exclude any possible monocyte contamination, as we[29] and others[45,48,104] did.

In contrast to the human system, there is no disagreement on the ability of murine neutrophils to produce IL-6, both in vitro and in vivo.[107] With regard to the in vitro system, Romani and colleagues (personal communication) reported that neutrophils purified from the peritoneal cavity of mice, and cultured in vitro with IFNγ plus LPS, or with different strains of *C.albicans* cells, express IL-6 messages (as revealed by RT-PCR), and produce substantial amounts of immunogenic IL-6, as measured by ELISA. The other in vivo findings will be widely discussed in chapter 8.

INTERLEUKIN-8 (IL-8)

Since the early demonstrations that neutrophils themselves can produce IL-8,[108,109] a large number of studies followed and extended such interesting findings. Strieter et al[108] provided the first molecular demonstration that neutrophils express IL-8. They showed that PMN (>99% pure) plated at 5×10^6/ml expressed significant steady-state levels of IL-8 mRNA after 4 h of stimulation with LPS (100 ng/ml) and released about 100 pg/ml of IL-8 into the culture supernatants after 24 h.[108] Almost concomitantly, we obtained very similar results with LPS (1 µg/ml),[109] and further showed that phagocytosis of Y-IgG represented a much more potent stimulus for the extracellular production of IL-8 release (from 10 to 50 ng/ml/18 h/10^7 PMN) (Fig. 3.2). In addition, we provided clear molecular and immunological evidence that the IL-8 recovered in the PMN cultures was not due to contaminating monocytes, and that it was biologically active.[109] Interestingly, a recent study showed that LPS-induced IL-8 production by neutrophils is enhanced by PMN binding to solid phase fibrinogen.[110]

Subsequently,[111] we were able to demonstrate that fMLP, a neutrophil chemotactic agonist, also induces PMN to release substantial amounts of IL-8 (0.5-2.0 ng/ml/3 h). Contrary to the Y-IgG- or LPS-stimulated release of IL-8, which is sustained over time, secretion of IL-8 in response to fMLP was transient, since maximal production was observed at 2-3 h and then returned to basal levels (see also Figs. 3.2 and 4.4).[111] This could be explained by the fact that IL-8, once produced, is taken up by cell surface receptors, or is degraded by proteolytic enzymes released by PMN.[112] This fMLP-induced IL-8 production was preceded by an enhanced expression of IL-8 mRNA (see also Fig. 3.3) and was dependent on de novo protein synthesis.[111] Furthermore, optimal release of IL-8 by fMLP was obtained at 10 nM, a concentration unable to stimulate a respiratory burst under identical experimental conditions, and was potentiated from 2- to 6-fold after a brief preincubation of PMN with cytochalasin B, a drug known to optimize neutrophil degranulation.[14] We also found that other neutrophil chemotactic factors, such as the complement component, C5a, and platelet-activating factor (PAF), were able to trigger the release of IL-8 by human PMN. The latter results were later substantially confirmed for fMLP[106] and C5a,[113] and extended to another classic PMN chemotactic factor, leukotriene B_4 (LTB_4).[114] It should be pointed out that in another work made by Strieter et al,[115] graded concentrations of C5a, fMLP, and LTB_4 apparently did not induce any significant extracellular antigenic IL-8, relative to unstimulated cells. Similar results were also reported by Au et al[116] with fMLP and PAF when used at 1 mM. However, in the two latter studies, extracellular IL-8 was measured in supernatants harvested 24 h post-stimulation. As explained above (and shown in Fig 3.2), the 24 h time point is not optimal to find IL-8 in supernatants from fMLP-stimulated PMN.[111] Strieter et al[115] nevertheless reported that if neutrophils were exposed to either of those chemotactic agonists in the presence of LPS, the production of IL-8 was synergistically elevated. The fact that chemotactic factors can induce the production of another chemoattractant (i.e., IL-8) is intriguing, especially considering that the release of other cytokines (such as TNFα, IL-1 or IL-12) by PMN in response to fMLP is undetectable (our unpublished observations and ref. 77). IL-8 production induced by chemotactic factors could represent a feedback mechanism whereby greater numbers of PMN are attracted to an inflammatory site, ensuring a persistent influx of PMN that perpetuate the inflammatory reaction. IL-8 is likely the main cause of the sustained local accumulation of neutrophils, due to its long-lasting effect.[117] By contrast, the other chemoattractants act more transiently because they are eventually inactivated by oxidation, hydrolysis or enzymatic cleavage.[118]

Strieter et al[115] also demonstrated that other than LPS, PMN stimulated with TNFα and IL-1β expressed IL-8 mRNA in a time- and dose-dependent manner, the order of potency being LPS > IL-1β > TNFα. Fujishma et al[119] also analyzed IL-8 gene expression and secretion in neutrophils stimulated with LPS, TNFα and IL-1β, and found that IL-1β (1000 U/ml) induced a lower level of mRNA expression and a smaller amount of IL-8 than LPS (100 ng/ml) and TNFα (1000 U/ml), after 24 h of incubation. We have later confirmed that IL-1β, over a wide concentration range, is a very weak stimulus for the extracellular production of IL-8.[120] However, we also observed that IL-1β dramatically synergizes with TNFα, even if the latter is used at the same doses produced by neutrophils in response to LPS.[120] Figure 4.2 shows a typical Northern blot analysis representing a time-course of IL-8 mRNA expression in neutrophils stimulated with TNFα.

The range of stimuli shown to induce the production of IL-8 has steadily increased during the last years (Table 4.6). For example, sulfatides have been shown to induce IL-8 mRNA in cultured PMN;[50] the possible implications of these observations have been already discussed in the TNFα section. Other studies identified GM-CSF as a very potent

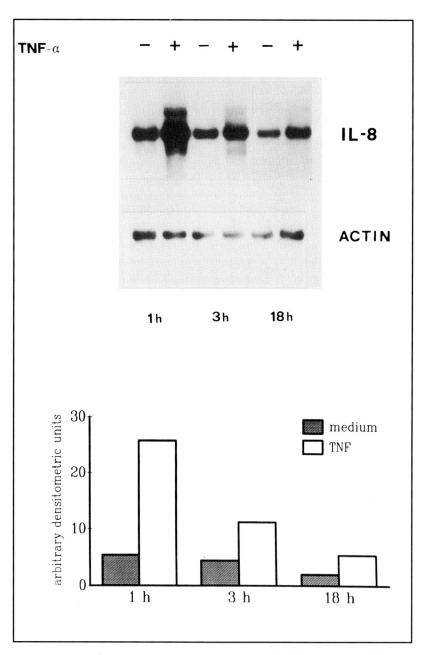

Fig. 4.2. Kinetics of IL-8 mRNA expression in TNFα-stimulated human neutrophils. PMN were cultured with TNFα (5 ng/ml) for the times indicated and Northern blot analyses were performed. The lower part of the figure depicts the related densitometric analysis.

Table 4.6. Effect of various agents on the production of IL-8 by human neutrophils

Agent	Effect (ref.)	
	Undetected	Stimulating
LPS	121	108–111, 115, 119, 123, 125, 129
fMLP	115, 116	106, 111, 125, 129
C5a	115	111, 113, 129
Leukotriene B$_4$	115	114
PAF	116	111
TNFα	121	115, 119, 120, 123, 125, 139
IL-1β	121	115, 119, 120
IL-2	**	124
IL-4	48, 107	
IL-8	119	109*, 124*
IL-10	48, 120, 157	
GM-CSF		121, 122, 124*, 139
G-CSF, IL-3	122, 180	
TGFβ	48	123
IFNγ	49, 125	
Prostaglandin E$_2$	107	
Substance P		90
bacterial toxins		136, 137
P. aeruginosa		138
C. albicans		124*
Y-IgG		30, 109, 125
Zymosan		116
RSV		106
PMA		106, 119, 129, 130*
A23187		106, 116, 129
Staurosporine		130
Dexametasone	107, 130*	
anti-CD11/CD18	116	
MSU/CPPD		139
Sulfatides		50*
CHX		124*, 125*

PAF: platelet activating factor.
* : mRNA only.
** : our unpublished observations.

inducer of IL-8 by human PMN. In the first one, McCain et al[121] indeed showed that GM-CSF (5 ng/ml) potently induced IL-8 gene expression, antigenic production and IL-8-dependent chemotactic activities, with maximal expression at 4 h. But surprisingly, neither LPS nor TNFα and IL-1 had significant effects towards the release of chemotactic activities into the media after 4 h.[121] Similarly, Takahashi et al[122] investigated the potential ability of G-CSF, IL-3 and GM-CSF to induce the production of IL-8 from PMN. Only GM-CSF was found to induce the synthesis and release of IL-8, with a time course very similar to that reported by Cain et al.[121] In addition to GM-CSF, TGFβ was reported to induce IL-8 mRNA and the release of small amounts of IL-8 in PMN supernatants,[123] but this finding has not been confirmed.[48] Finally, Wei et al[124] showed that GM-CSF, IL-2, and heat-killed *C.albicans* all induce IL-8 in PMN. In particular, IL-2 enhanced both IL-8 mRNA transcripts and IL-8 extracellular production, the latter reaching maximal levels by 4-6 h of stimulation with 1000 U/ml (which corresponded to the optimal doses). Supernatants from IL-2-stimulated PMN exhibited a potent chemotactic activity towards freshly isolated PMN that could be specifically blocked by anti-IL-8 monoclonal antibodies. Furthermore, the increase of IL-8 mRNA by IL-2 observed at its peak was unaffected by CHX (10 μg/ml), suggesting that new protein synthesis was not required for IL-2-induced IL-8 mRNA. In contrast, CHX decreased by almost 95% the levels of IL-8 mRNA induced by GM-CSF (1000 U/ml), whereas it superinduced the levels of IL-8 mRNA in response to IL-8, indicating that modulation of IL-8 mRNA levels occurs through several different pathways, depending on the agonist used. In the same study, a 4 h treatment of PMN with CHX alone did not result in any particular effect on IL-8 mRNA; in contrast, we reported that CHX superinduces IL-8 mRNA in PMN.[125] Furthermore, as already discussed in the case of the supposed TNFα production induced by IL-2,[36] we were unable to detect any increase of IL-8 mRNA or protein secretion in neutrophils stimulated with IL-2.

Exposure of neutrophils (>98% pure) to respiratory syncytial virus (RSV) particles provoked an enhancement of IL-8 mRNA steady-state levels, accompanied by the secretion of IL-8, in a time- and dose-dependent manner.[106] Synthesis of IL-8 from PMN depended on the adherence of viral particles as well as on a phagocytic event and not on the infection process, as demonstrated by the fact that infectious RSV particles and UV-inactivated RSV particles were equally effective.[106] Interestingly, if the RSV particles were opsonized with monoclonal antibodies directed to the RSV-fusion protein (F protein), IL-8 release from PMN increased (in comparison with RSV alone), and this increase occurred without a concomitant enhancement of IL-8 mRNA levels. These results are highly reminiscent of those obtained by us using Y and Y-IgG as stimuli.[30] In spite of an equivalent increase of IL-8 mRNA levels induced by both Y-IgG and Y,[30] Y-IgG was more potent than Y in inducing the release of IL-8.[30] Involvement of the Fcγ-receptors by Y-IgG might play role in enhancing the synthesis and/or secretion rate of the de novo-synthesized cytoplasmic IL-8 pool (see chapters 6 and 9 for a discussion on the ability of Y-IgG to act at the level of IL-8 secretion).

As demonstrated by Au et al,[116] coincubation of neutrophils with zymosan (which is a yeast cell wall extract) also resulted in the appearance of immunoreactive IL-8 in the PMN supernatants, which was detectable at 8 h, and reached a maximum at 24 h. Zymosan-stimulated IL-8 production was dose-dependent and was abolished by protein synthesis inhibitors.[116] Furthermore, confirming our previous observations made with unopsonized yeasts particles,[30] Au et al showed that opsonization of zymosan was not a prerequisite for IL-8 production. As binding of neutrophils to opsonized or unopsonized zymosan has been shown to involve complement receptor CR3(CD11b/CD18),[126] the authors investigated whether stimulation of this adhesion receptor complex was involved in inducing IL-8 synthesis. They showed that addition of anti-CD18 monoclonal antibodies virtually abolished, and addition of anti-CD11b monoclonal antibodies substantially inhibited, the IL-8 production by neutrophils stimulated by zymosan, whereas monoclonal antibodies alone did not induce a detectable cytokine release. Since PAF,[127] IL-1[45,65] and, likely, TNFα are secreted by zymosan-stimulated neutrophils, the authors also examined whether these mediators could play a role in the zymosan-induced production of IL-8. Neither addition of human IL-1ra, nor anti-human TNFα monoclonal antibodies had any effect, whereas two structurally distinct PAF antagonists suppressed zymosan-induced neutrophil IL-8 generation by approximately 70%. These observations suggested that IL-8 release from neutrophils stimulated by zymosan is dependent on a CD11b/CD18 signaling pathway, and that the transduction mechanisms involved appear to be regulated by endogenously generated PAF or by related molecule(s).

Au et al[116] also observed that the calcium ionophore A23187 (10^{-7} M) induced a marked IL-8 generation from neutrophils, over a 24 h time course. Interestingly, A23187 was partially inhibited if co-administered with a selective PAF antagonist, in line with the known capacity of A23187 to stimulate the synthesis of PAF in neutrophils.[128] That A23187 represents an efficient and powerful stimulus for IL-8 synthesis and release from PMN has been recently confirmed by Arnold et al,[106] and by Kuhns and Gallin.[129] The latter authors showed that after an incubation of 8 h with 1 µM A23187, cell-associated IL-8 increased over 100-fold and more than 35% of it was released. Other than confirming that fMLP, C5a and LPS induce new synthesis and secretion of IL-8 from PMN, Kuhns and Gallin[129] indicated also PMA (20 ng/ml) as a good IL-8 secretagogue, thereby extending the previous observations made by Fujishima et al,[119] by us,[130] and by Arnold et al[106] on the ability of PMA to increase IL-8 mRNA steady-state levels. In agreement with our earlier data,[47] Kuhns and Gallin[129] also found that in PMN, cell-associated IL-8 increases several-fold during incubation at 37°C in vitro, and much more (200-fold) after treatment with stimuli. CHX, but not actinomycin D, inhibited the accumulation of cell-associated IL-8 after culture at 37°C, suggesting that accumulation of IL-8 was under translational rather than transcriptional control. Finally, by using sucrose density centrifugation of freshly isolated neutrophils disrupted by nitrogen cavitation, Kuhns and Gallin[129] were able to localize a portion of IL-8 to a subcellular fraction of heterogenous, light membranous organelles. The latter, originally identified by Borregaard et al,[131] coeluted with latent alkaline phosphatase, and were distinct from azurophilic and specific granules. To date, this is one of the few works that identify the subcellular localization of a cytokine in human neutrophils.[129]

Serra et al,[90] continuing to add new stimuli to the list of the inducers of IL-8, identified the neuropeptide, substance P (SP), one of the main mediators of neurogenic inflammation,[132] as able to directly stimulate the release of IL-8 from human neutrophils. Induction of IL-8 release by SP occurred in 2 h and was preceded by an increase of IL-8 mRNA steady-state levels. Remarkably, SP also enhanced the IL-8 release induced by TNFα and fMLP.[90] It was already known that SP acts on PMN, inducing chemotaxis, potentiating phagocytosis, stimulating respiratory burst, exocytosis, antibody-dependent cell-mediated cytotoxicity, and priming them.[90,133,134] The ability of SP to induce long-term changes in PMN, in addition to the rapid effects on the activation state of these cells, is relevant to the role played by sensory nerves in the modulation of the inflammatory response.

Very surprisingly, staurosporine (a nonspecific inhibitor of protein kinase C and of other kinases[135]), was also reported to induce, in a dose- and time-dependent fashion, IL-8 mRNA and protein secretion by human PMN.[130] The biological meaning and the physiological implications of all these effects of staurosporine are unknown and remain to be clarified.

Many different bacterial toxins can activate PMN to express IL-8 mRNA and protein release.[136] This was true for the two pore-forming toxins, Panton-Valentine leukocidin (Luk-PV)(from *S.aureus* V8), and alveolysin (Alv)(from *Bacillus alvei*), as well as for the erythrogenic toxin A (ETA)(from *Streptococcus pyogenes*).[136] Production of IL-8 occurred only when all toxins were used at low concentrations (0.5-5 ng/10^7 cells), over a 16 h period.[136] Different pathways mediated the generation and release of IL-8 induced by these toxins, since inhibitors of protein tyrosine kinases reduced the release and the mRNA expression of IL-8 in PMN challenged with Luk-PV and Alv, but not with ETA.[136] Furthermore, a very low molecular weight (1 kD) product of *Pseudomonas aeruginosa* was shown to induce the expression of IL-8 mRNA in neutrophils.[137] In this regard, clinical *P.aeruginosa* isolates, the mucoid

P.aeruginosa strain (CF3M) and its nonmucoid revertant (CF3), and purified *P.aeruginosa* mucoid exopolysaccharide (alginate), were all shown to produce a significant increase in IL-8 release from human PMN, in a dose- and time-dependent manner.[138] These results have particularly important implications in the context of cystic fibrosis (CF), since they not only attribute an important role for PMN-derived IL-8 in maintaining neutrophil influx, but provide a better understanding of the mechanisms by which *P.aeruginosa* bacteria determine the clinical outcome of CF and other diseases.[138]

Lastly, Hachicha et al[139] showed that the inflammatory microcrystals MSU and CPPD, the major mediators of gout and pseudogout, respectively, both increased the secretion of IL-8 by neutrophils in a dose- and time-dependent manner, but had no effect on that of MIP-1α. Importantly, the presence of MSU and CPPD synergistically enhanced the production of IL-8 induced by TNFα and GM-CSF, but completely inhibited MIP-1α secretion induced by TNFα.[139] The implications of all these latter findings are discussed in chapter 7, whereas the relevance of the ability of neutrophils to produce IL-8 can be deducted by the numerous actions of this chemokine, mainly towards neutrophils themselves, already described in chapter 2.

On a final note, it is also correct to mention here that in contrast to the data of Strieter et al,[115] we,[111] as well as others,[119,138] observed the presence of IL-8 mRNA in most samples of freshly isolated neutrophils. Usually, these constitutive IL-8 transcripts decrease almost completely within a few hours of cell culture in the absence of stimulation.[111] Despite the presence of specific mRNA, secretion of IL-8 by unstimulated cultured human neutrophils is always very low (below 100 pg/ml), unless neutrophils have been inadvertently preactivated by the isolation procedures. Similarly, other groups have observed the presence of IL-8 mRNA in resting cultured PMN.[106,115,119,122,124] Although adherence to plastic has been reported to induce monocyte and neutrophil secretion of IL-8,[115,140] we[111] and others[122] did not find substantial differences in culturing neutrophils under adherent (polystyrene) or nonadherent (polypropylene) conditions. We do not know whether the constitutive IL-8 mRNA results from mechanical stress during cell preparation, or represents a constitutive RNA pool that may facilitate the appearance of the mature protein when the PMN are stimulated. Evidence for the latter option is that the constitutive presence of the other cytokines mRNA, TNFα[111] or IL-1β[119] in freshly isolated PMN is much more variable compared to that of IL-8 mRNA. However, it cannot be excluded that cell separation procedures induce IL-8 but not TNFα or IL-1β mRNA expression.

INTERLEUKIN-12 (IL-12)

The only demonstration that mature human neutrophils produce and release both the IL-12p40 free chain and the IL-12p70 heterodimer is that published by my group.[141] However, our in vitro findings have been recently corroborated by elegant in vitro and in vivo observations made by Romani et al with murine neutrophils (personal communication), which are described later in this paragraph and in chapter 8. Our experiments showed that, among a wide range of stimuli tested, including TNFα, fMLP, PMA and Y-IgG, only LPS in combination with IFNγ efficiently induced a significant production of biologically active IL-12. This was because on the one hand, LPS induced a 100-fold increase in IL-12p40 mRNA without having an effect on IL-12p35 mRNA accumulation; on the other, IFNγ directly induced a several-fold increase in the accumulation of IL-12p35 mRNA, and enhanced the LPS-induced accumulation of IL-12p40 mRNA. Recently, a similar molecular regulation of IL-12 has been reported to occur in murine RAW264.7 macrophage cells.[142] It should be pointed out that, by Northern blotting, we could detect only IL-12p40 transcripts, but not IL-12p35; the presence of the latter transcripts were instead measured by the more sensitive and quantitative ribonuclease protection assay.[141] Therefore, the combined effect of LPS and IFNγ induced sufficient expression of both IL-12p40 and IL-12p35 mRNAs to attain production of the biologically active IL-12p70 heterodimer at physiologically relevant concentrations.[141] The kinetics of LPS plus IFNγ-induced IL-12p40 and IL-12p70 production by PMN were relatively delayed (peaking at 48 h) compared to the productions of TNFα, IL-1β, IL-8 and IL-1ra. Whether the LPS plus IFNγ-induction of IL-12 is an indirect result of the stimulation of the cytokines endogenously produced by PMN under these conditions remains to be established. As already observed in monocytes/macrophages,[143] IL-10 suppressed the LPS-induced IL-12p40 mRNA (see also Fig. 5.4) and protein secretion in PMN.[141] Furthermore, Figure 4.3 shows differences between PMN and monocytes (again purified from the same donor) regarding their ability to produce IL-12p40. It is evident that in PMN, phagocytosis is ineffective in inducing IL-12p40 production[141], whereas in monocytes it represents a stimulus even more powerful than LPS, in agreement with published data.[144]

Interestingly, Hayes et al,[145] recently confirmed that the genes encoding the two IL-12 subunits are independently and finely regulated. They in fact showed that in human monocytes isolated by elutriation, IL-12p40 and IL-12p35 mRNAs were directly inducible by LPS, but optimal induction of both genes occurred only after a preincubation with IFNγ, that is after priming. Kinetics of priming and induction of IL-12p40 and IL-12p35 were distinct, in that for example, priming by IFNγ for IL-12p40 was transient, whereas for IL-12p35 was not. While both IFNγ and GM-CSF, but not M-CSF, provided an effective priming for IL-12p40 induction, LPS-induction of IL-12p35 was specifically dependent on IFNγ. Moreover, as we observed in neutrophils,[141] the level of active IL-12 protein in monocytes was controlled

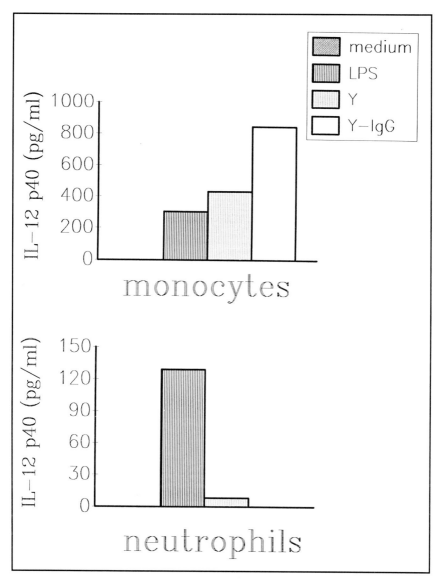

Fig. 4.3. Comparison of the ability of neutrophils and monocytes to produce IL-12p40. PMN (5×10^6/ml) and monocytes (5×10^5/ml) isolated from the same donor were cultured for 24 h with LPS or S.cerevisiae [either alone (Y) or opsonized with IgG (Y-IgG)], before measuring IL-12p40 in the cell-free supernatants.

by the regulated expression of IL-12p35.[145] In fact, in monocytes treated with LPS alone minimal IL-12 activity was released, while IFNγ-primed monocytes subsequently stimulated with LPS displayed strikingly high levels of IL-12.[145]

Very recently, Romani and colleagues (personal communication) purified neutrophils from the peritoneal cavity of naive mice injected 18 h earlier with 1 ml of thioglycollate using immunomagnetic beads, and cultured them in the presence of either IFNγ plus LPS or *C.albicans* cells for 24 h. The *C.albicans* cells utilized were the agerminative live vaccine strain PCA-2, which provokes healing infections, and the highly virulent CA-6 strain, which induces nonhealing infections. Under these conditions, IL-6 and TNFα, but not IL-4, could be detected in response to the different stimuli, including *C.albicans*. Interestingly, in PMN cultures stimulated with IFNγ plus LPS, not only low levels of bioactive IL-12 were found, in accord with our observations,[141] but also IL-10. IL-12, but not IL-10 was produced in response to the agerminative strain PCA-2, while the opposite pattern occurred in response to CA-6. Not only a similar profile of cytokine production was observed in neutrophils from peripheral blood, but, as revealed by RT-PCR, both IL-10 and IL-12p40 messages, together with those of TNFα and IL-6 were present in neutrophils obtained from the peritoneal cavity and cultured with LPS plus IFNγ for 2 h. In contrast, a similar analysis of peritoneal macrophages revealed the presence of IL-6 and TNFα messages, but not of IL-10 and IL-12p40. Furthermore, by an immunophenotype and fluorescence microscopic analyses, the presence of IL-10 and IL-12 in neutrophils was documented. Altogether, these data indicates that murine neutrophils are endowed with the ability to produce IL-12 (and IL-10) in response to different stimuli in vitro, including *C.albicans* cells. The ability of neutrophils to produce IL-10 and IL-12 in vivo, and the meaning and importance of such function in the context of mounting T helper 1 (Th1)/Th2 response, will be described in chapter 8.

Because of the important immunoregulatory functions of IL-12, in particular the induction of IFNγ production and facilitation of Th1 type responses, the ability of neutrophils to produce IL-12 (see Table 4.7 for a summary) suggests that they may play an active role in the regulatory interactions between innate resistance and adaptive immunity. Production of IL-12, TNFα, IL-1β, and other molecules by PMN, and of IFNγ by NK cells, well known to affect generation of Th1 cells and other aspects of T and B cell functions, suggests that the activation of innate resistance early during infection represents not only a first line of defense against the invading microorganisms, but also that it could prime the immune system and determine the characteristics of the ensuing antigen-specific adaptive immune response. It is possible to imagine a dynamic situation in which IL-12 produced by PMN (and other phagocytic cells) in response to bacterial or parasitic infec-

Table 4.7. Effect of various agents on the production of IL-12 by human neutrophils

Agent	Effect (ref.)	
	Irrelevant	Inducing
LPS		141*
fMLP	141	
TNFα	141	
IL-10	141	
IFNγ		141**
IFNγ + LPS		141
C. albicans		***
Y-IgG	141	
PMA	141	

* : only IL-12p40 mRNA and protein.
** : only IL-12p35 mRNA.
*** : L. Romani, personal communication.

tions, induces the production of IFNγ by T and NK cells, which in turn activates the phagocytic cells,[146] and at the same time favors a Th1-type immune response. If so, IL-12 could be involved in an effective positive feedback mechanism that, through IFNγ, could induce the activation of PMN and monocytes. As such, IL-12 could be recognized as a bridge between innate resistance and adaptive immunity.

GROWTH RELATED GENE PRODUCT-α (GROα) AND GROβ

Haskill and coworkers firstly demonstrated that human neutrophils could express GROα mRNA.[147] By PCR analysis, they showed that PMN which have been adhered for 45 min to fibronectin-coated plastic selectively expressed GROα mRNA, whereas, under similar experimental conditions, monocytes from the same individual expressed all three GRO isoforms (GROα, GROβ and GROγ).[147] Iida et al[148], in a different study, showed by Northern blot analysis that PMN constitutively express also GROβ mRNA at very low levels.[148] A 5-10-fold induction of these GROβ transcripts were however observed upon stimulation with LPS (1 µg/ml) for 24 h, whereas in monocytes the same treatment resulted in a more than a 100-fold GROβ mRNA increase.[148] No investigations on the GROα and GROβ synthesis were performed by those authors,[147,148] most likely because of the lack of available specific assays to measure the two chemokines at the time.

In one of our most recent studies,[149] we examined the possibility that PMN might secrete GROα, by using a new, specific ELISA. We found that activated human PMN are able to release GROα, in the order of picograms/ml range.[149] In our hands, phagocytosis of Y-IgG was once again the most powerful inducer of GROα, being approximately 2- to 3-fold more potent than LPS (1 µg/ml), and much more than TNFα (5 ng/ml). This order of potency was in fact similar to that previously observed for studying the production of IL-8 and TNFα in PMN.[29,30,109,120] By contrast, LPS was more potent than Y-IgG, TNFα, and fMLP in inducing the production of GROα by monocytes, indicating that the regulation of GROα production is governed by distinct mechanisms in both cell types.[149] The release of GROα by stimulated PMN was preceded by an accumulation of GROα mRNA, whose kinetics varied depending on the agonist.[149] An exception was observed with fMLP however. Very small quantities of GROα protein were de-

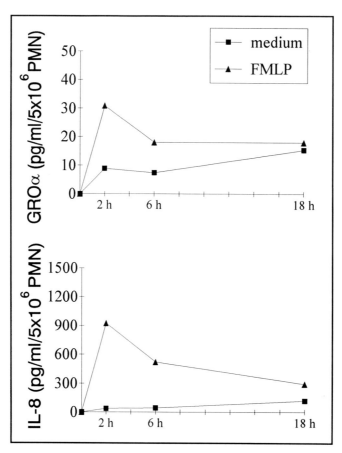

Fig. 4.4. Comparison of the neutrophil ability to release GROα and IL-8 in response to fMLP.

tected in the cell-free supernatants of fMLP-stimulated PMN (Fig. 4.4), in spite of a very high accumulation of GROα mRNA transcripts that reached levels comparable to those induced by LPS.[149] In this regard, it is worth recalling that fMLP is a good inducer of IL-8 release[111] (see Figs. 3.2, and 4.4). We do not know whether GROα is actually released by PMN upon fMLP stimulation, but then rapidly degraded by the proteolytic enzymes simultaneously secreted in those conditions. It also cannot be excluded that in spite of its ability to induce GROα mRNA, fMLP fails to provide the intracellular signals necessary to either translate or secrete GROα. If the latter hypothesis is true, it would suggest that GROα production in PMN is controlled at the translational or post-translational level, similar to what has been observed, for example, during malignant transformation of normal melanocytes.[150] Our results with Y-IgG would also support a translational or post-translational regulatory mechanism controlling GROα production. As already mentioned, Y-IgG induced the highest release of GROα into the neutrophil culture supernatants, but only caused a moderate GROα mRNA accumulation, compared to LPS, TNFα, or fMLP.[149] Interestingly, and in keeping with a possible control at the level of GROα secretion by Y-IgG, we recently showed that the efficiency of IL-8 release in Y-IgG-stimulated PMN was much higher than in LPS- or TNFα-activated cells.[47] As will be discussed in chapter 5, our study also showed that the extracellular production of GROα by PMN can be modulated by IFNγ and IL-10.[149]

Two recent works recognized neutrophils as cells potentially producing GROα, substantially confirming our results. Koch et al[151] examined whether PMN obtained from peripheral blood (PB) or from RA SF generated GROα. They found that after a 24 h culture, both RA SF PMN and normal PB PMN constitutively produced very high levels of GROα (about 3.5 ng/ml/10^7 cells), and that this production was further increased upon stimulation with 1 µg/ml LPS.[151] Hachicha et al[139] assessed the levels of GROα after stimulation of neutrophils for up to 12 h with TNFα, GM-CSF, MSU and CPPD, used either individually or in combination. Low levels (in the order of picogram/ml) of GROα were detected in both control and stimulated cell supernatants, regardless of the triggering conditions.[139] Although the number of PMN used in these latter experiments were not specified, the quantitative results shown with TNFα are essentially in agreement with ours.

The ability of PMN to produce GROα is relevant for several reasons. The molecular basis of the recruitment of neutrophils, monocytes, lymphocytes and other leukocytes to sites of inflammation, a process critical for host defense after bacterial infection or tissue injury, is not yet fully elucidated. Similar to IL-8, GROα acts as a mediator of inflammation, as it induces chemotaxis, exocytosis and enhancement of the expression of CD11b/CD18 and CR1 in neutrophils.[152-154] Thus, in addition to the production of IL-8, the generation of GROα by neutrophils may contribute to stimulate the recruitment to, and activation of further neutrophils at sites of inflammation. Although still too early to speculate about the in vivo role of neutrophil-derived GROα, the expression of KC mRNA, which is the murine analogue of GROα,[117] was recently found within exudate neutrophils obtained after intraperitoneal injection of thioglycollate in rats.[155] Moreover, the fact that among its various biological activities, GROα possess mitogenic effects on normal and transformed human melanocyte cell lines,[117] raises the possibility that neutrophil-derived GROα might play a role in melanocyte transformation. Table 4.8 schematically lists the experimental conditions utilized thus far to detect the production of GROα by neutrophils.

MACROPHAGE INFLAMMATORY PROTEIN-1α AND -1β (MIP-1α/β)

A very important advance in the characterization of the cytokines produced by PMN, has stemmed from the series of studies made by Kasama and coworkers,[156,157] who identified MIP-1α and MIP-1β as being released by these cells. As outlined in chapter 2, MIP-1α and MIP-1β are members of the β-chemokines (C-C subfamily) of the chemokine supergene family, which act as potent chemotactic/activating factors for monocytes and subpopulations of T lymphocytes, and which also activate several effector functions of macrophages and neutrophils.[117] In their first work, Kasama et al[156] observed that cell-free supernatants from human PMN stimulated with increasing concentra-

Table 4.8. Effect of various agents on the production of GROα by human neutrophils

Agent	Effect (ref.)	
	Undetected	Stimulating
Medium		139, 151
LPS		149, 151
fMLP		149
TNFα		139, 149
IL-10	149	
GM-CSF		139
IFNγ	149	
Y-IgG		149
Fibronectin		147
MSU/CPPD		139

tions of LPS (1-100 ng/ml) possessed a dose-dependent chemotactic activity for human monocytes, which was significantly attenuated (by approximately 60%) in the presence of neutralizing anti-human MIP-1α antibodies. By using Northern blot analysis, ELISA and immunocytochemistry, and excluding the potential contribution of possible contaminating monocytes, they were able to demonstrate that LPS induced a dose- and time-dependent mRNA expression and extracellular production of both MIP-1α,[156] and subsequently, of MIP-1β[157] as well, from PMN. While GM-CSF, G-CSF and IL-3 failed by themselves to induce the expression of MIP-1α over a wide concentration range, stimulation of PMN in the presence of both LPS and GM-CSF resulted in a synergistic expression of both MIP-1α mRNA and protein, compared with LPS alone.[156] One of the mechanisms involved in the increase in MIP-1α production by GM-CSF plus LPS-treated PMN appeared to be the stabilization of MIP-1a mRNA, which resulted in a prolonged MIP-1α mRNA half-life.[156] Actions at the level of half-life of both MIP-1α and MIP-1β mRNAs in LPS-treated PMN were also exerted by IFNγ and IL-10,[49,157] but these aspects will be discussed in greater detail in chapter 5.

Recently, Hachicha and colleagues extended to other stimuli the ability of inducing neutrophil secretion of MIP-1α, and also showed that MIP-1α expression is regulated differently than that of IL-8.[139] Among MSU, CPPD, GM-CSF and TNFα, only the latter exerted a significant effect on MIP-1α mRNA expression and secretion in neutrophils.[139] In this regard, Kasama et al[49] also provided indirect evidence that endogenous TNFα was responsible for MIP-1α and MIP-1β production by LPS-stimulated PMN, by using neutralizing anti-TNFα antibodies. But the major finding of potential biological significance in the studies by Hachicha and coworkers was that while secretion of MIP-1α induced by TNFα was completely inhibited in the presence of either MSU or CPPD, the production of IL-8 in the same conditions was synergistically enhanced. These results might imply that the failure of inflammatory microcrystals to directly induce MIP-1α production (while inducing that of IL-8),[139] as well as their ability to inhibit the production of MIP-1α in response to TNFα, may prevent the generation of a chemotactic signal by neutrophils that could potentially attract a large numbers of mononuclear cells to the synovial environment. Strikingly, the effect of combining TNFα with either MSU or CPPD did not inhibit the expression of MIP-1α mRNA induced by TNFα, but rather enhanced it,[139] suggesting that the inhibitory effect was mediated at the translational level. However, it must be considered that Northern blot experiments were performed only at a single time point (3 h), while MIP-1α secretion was measured after 12 h. Finally, our unpublished experiments confirmed that neutrophils express MIP-1α and MIP-1β mRNAs in response to both LPS and

TNFα. We also found that MIP-1α mRNA can be induced by Y-IgG, but not by fMLP or IL-8.

The reported ability of PMN to secrete MIP-1α (see Table 4.9 for a summary) and MIP-1β is of considerable importance. The fact that PMN also produce IL-8 and GROα, suggests that granulocytes, once arrived at the inflammatory site, can promote not only the further recruitment of neutrophils, but also the subsequent accumulation and activation of monocytes/macrophages and lymphocytes. Thus, it appears that PMN, through the production of IL-8, GROα, MIP-1α, MIP-1β and IP-10 (see below), could play a pivotal role in regulating the switch of the type of leukocyte infiltration typically observed during the evolution of the inflammatory response from the acute to chronic stage.

MONOCYTE CHEMOTACTIC PROTEINS (MCP-1, MCP-2 AND MCP-3)

Attempts to define whether MCP-1 mRNA expression occurs in adherent or LPS-stimulated neutrophils were performed by Strieter et

Table 4.9. Effect of various agents on the production of MIP–1α and MCP-1 by human neutrophils

Agent	Effect on MIP–1α (ref.)		Effect on MCP–1 (ref.)	
	Undetected	Stimulating	Undetected	Stimulating
Medium			108*	159
LPS		156, 157, **	108*, 158	
fMLP	**			
TNFα		139, **	139	
IL–1β			158	
IL–3	156			
IL–8	**			
IL–10	157, **			
G–CSF	156			
GM–CSF	139, 156		139, 158	159
IFNβ			158	
IFNγ	49		158	
Y–IgG		**		
PMA			158	
retinoic acid				159
MSU/CPPD	139		139	

* : mRNA only.
** : our unpublished observations.

al.[108] However, in contrast to the very high expression of IL-8 mRNA after 4 and 8 h of culture with LPS, MCP-1 mRNA was absent.[108] Presumably discouraged by these negative results, Strieter et al did not attempt to detect MCP-1 protein.[108] That was done a few years later by Van Damme et al,[158] who had developed sensitive radioimmunoassays not only for MCP-1, but also for MCP-2. In agreement with Strieter's study,[108] Van Damme et al[158] did not detect any extracellular release of either MCP-1 or MCP-2 by granulocytes, even though the cells were stimulated for up to 48 h with optimal doses of many agonists, including IL-1β, IFNβ, IFNγ, GM-CSF, PMA, and LPS.[158]

In a recent study, however, Burn et al[159] reported that MCP-1 transcripts were readily detectable in neutrophils incubated for 20 h in culture medium containing 10% endotoxin-free fetal calf serum for 20 h, as measured by Northern analysis or RT-PCR, but were absent in freshly-isolated neutrophils.[159] Furthermore, by immunocytochemistry, the same authors showed that PMN incubated for 20 h either in medium alone or with GM-CSF, stained positively for the MCP-1 protein.[159] However, MCP-1 protein could not be detected in freshly isolated cells.[159] Although this is also discussed in chapter 8, it is worth mentioning here that in a rat model of bleomycin-induced lung injury, PMN stained positive for MCP-1,[160] whereas D'Angio et al[161] reported that in lavageable cells from lung of rabbits exposed to hyperoxia, recruited PMN expressed only IL-8, but not MCP-1 mRNA. Finally, Hachicha et al[139] recently reported that mRNAs for MCP-1, MCP-2, MCP-3, RANTES or I-309 were undetectable in unstimulated PMN, or in PMN stimulated for 3 h with TNFα, GM-CSF, MSU and CPPD, either alone or in combination.

As in the case of the IL-6 gene, further studies are necessary to clarify whether, and under which conditions, human neutrophils can clearly express MCP-1 mRNA and protein. Table 4.9 summarizes the experimental conditions used to date to examine the expression of MCP-1 by human neutrophils.

INTERFERON INDUCIBLE PROTEIN-10 (IP-10)

Ongoing studies in my laboratory suggest that human neutrophils are also able to produce and release IP-10. The only stimuli effective on PMN appear to be IFNγ in combination with either LPS or TNFα. Release of IP-10 is preceded by an inducible accumulation of IP-10 mRNA, which parallels the effects of the above mentioned stimuli. Interestingly, while IFNγ alone induces the expression of detectable levels of IP-10 mRNA (as measured by Northern blot analysis), no IP-10 protein is present into the cell-free supernatants of IFNγ-treated PMN. Under the same conditions, IFNγ alone seems to be a potent stimulus for IP-10 production by monocytes, suggesting, once again, a different molecular control of IP-10 production in neutrophils versus monocytes. Therefore, the recruitment of monocytes, lymphocytes and other leukocytes to sites of inflammation[161a] could be potentially regulated by neutrophils through the production of IP-10, in addition to MIP-1α and MIP-1β.

CD30 LIGAND (CD30L)

CD30, a member of the NGF/TNF receptor superfamily[162] is preferentially expressed by Th2-type CD4+ T cell clones,[163] as well as by CD8+ T cell clones showing a Th2 profile of cytokine secretion.[164] The natural ligand for CD30 (CD30L) has been cloned[165] and shown to belong to the superfamily of NGF/TNF ligands.[166] CD30L has been found to be expressed by B lymphocytes,[167] as well as by a subset of activated macrophages and T cells.[168] CD30 triggering induces nuclear factor-κB activation in human T cells,[169] and may provide a critical costimulatory signal for the development of Th2 responses.[170]

By Northern blot analysis, Gruss et al[171] observed that PMN constitutively express CD30L mRNA, but not that encoding CD30. Even though a 24 h treatment with LPS, IFNγ, or GM-CSF, slightly increased the steady-state levels of CD30L mRNA, Gruss et al[171] did not show any immunofluorescence analysis of CD30L surface expression on neutrophils. However, they recently detected the CD30L antigen in neutrophils present in specific hematologic malignancies (H.J. Gruss, personal communication). We also investigated whether human neutrophils can express CD30L. First, we confirmed that PMN express high levels of CD30L mRNA, especially after stimulation with LPS or fMLP. Second, by immunofluorescence analysis, we found a variable CD30L protein expression in neutrophils depending on the donor (our unpublished observations). Should CD30L protein expression in neutrophils be definitively ascertained, its biological implication would also need to be further investigated. For instance, it is known that the interaction between CD30L and CD30 triggers the replication of HIV.[172] Interestingly, Ho et al[173] showed that human neutrophils can potentiate HIV replication in infected mononuclear leukocytes, but did not identify the mechanisms or the molecules implicated in these effects. In the light of the potential ability of neutrophils to express CD30L, one obvious candidate for this interesting effect could be CD30L.

OTHER CYTOKINES

Isolated observations, listed below, would indicate that neutrophils can express or produce other cytokines, but I would like to stress that further studies are eagerly awaited to confirm these reports, in order to avoid premature speculations.

For example, Lindemann et al[20] reported that stimulation of PMN with GM-CSF (250 ng/ml) induces the mRNA expression and release of G-CSF and M-CSF, as determined by Northern analysis and in biological assays, respectively. Kita and coworkers,[174] on the basis of indirect effects detected in a biological assay, assumed that neutrophils stimulated with PMA in combination with ionomycin (a calcium ionophore), produced IL-3 and GM-CSF. They were, however, unable to directly measure by specific immunological techniques IL-3 and GM-CSF in neutrophil culture supernatants.[174] Similarly, Contrino et al[77] failed to detect the GM-CSF protein in cell lysates and supernatants from PMN (for 3 and 24 h) stimulated with TNFα, LPS, fMLP, or IL-1β.

Ramenghi et al[175] evaluated the cell types in peripheral blood that express the mRNA for stem cell factor (SCF), the ligand of the *c-kit* proto-oncogene.[176] SCF has been shown to play a critical role in the migration of melanocytes and germ cells during embryogenesis, as well as in the proliferative control of the hematopoietic compartment.[176] As revealed by RT-PCR, human granulocytes appeared to express both transcripts encoding the soluble and transmembrane forms of SCF.[175] Conversely, visible PCR products were not found in isolated lymphocytes, monocytes, and the total cell population obtained by Ficoll purification.[175] Ramenghi et al[175] did not investigate whether SCF mRNA expression was increased by stimulation of neutrophils, or whereas it was followed by synthesis of the related protein. Clearly, the latter experiments are mandatory before making any speculation on the possible biological significance of SCF expression by PMN.

In another report,[177] it has also been shown that granulocytes stimulated by fMLP plus cytochalasin B are able to release immunoreactive TGFα. Although it was demonstrated that eosinophils were the major source of TGFα, it could not be excluded that neutrophils might also produce the molecule. Again, confirmatory experiments are awaited.

Finally, as observed for RANTES, MCP-2, MCP-3, and I-309,[139] human neutrophils neither seem to produce any IL-2 activity,[68] nor IL-10 or IL-13 (M.A. Gougerot-Pocidalo, personal communication). In contrast, and as already mentioned above, Romani and coworkers (personal communication) reported that purified murine neutrophils can produce IL-10 in vitro, but not IL-4, in response to either IFNγ plus LPS, or to *C.albicans*.

REFERENCES

1. Wardlaw AJ, Moqbel R, Kay B. Eosinophils: biology and role in disease. Adv Immunol 1995; 60:151-266.
2. Warner JA, Kroegel C. Pulmonary immune cells in health and disease: mast cells and basophils. Eur Respir J 1994; 7:1326-1341.
3. Shirafuji N, Matsuda S, Ogura H et al. Granulocyte-Colony-stimulating factor stimulates human mature neutrophilic granulocytes to produce interferon-α. Blood 1990; 75:17-19.
4. Hiscott J, Ryals J, Dierks P et al. The expression of human interferon alpha genes. Philosophical transactions of the Royal Society of London. 1984; 307:217-226.
5. Rubinstein M, Levy WP, Morchera JA et al. Human leukocyte interferon:

isolation and characterization of several molecular forms. Arch Biochem Biophys 1981; 210:307-318.

6. Brandt ER, Linnane AW, Devenish RJ. Expression of IFNα genes in subpopulations of peripheral blood cells. Br J Hematol 1994; 86:717-725.
7. Tsuro S, Fuijsawa H, Taniguchi M et al. Mechanism of protection during the early phase of a generalised viral infection: contribution of polymorphonuclear leukocytes to protection against intravenous infection with influenza virus. J Gen Virol 1987; 68:419-424.
8. West BC, Eschete ML, Cox ME et al. Neutrophil uptake of vaccinia virus in vitro. J Infect Dis 1987; 156:597-606.
9. Klebanoff SJ, Coombs RW. Viricidal effect of polymorphonuclear leukocytes on human immunodeficiency virus-1. J Clin Invest 1992; 89:2014-2017.
10. Fujioka N, Akazawa N, Sakamoto K et al. Potential application of human interferon-α in microbial infections of the oral cavity. J Interferon Cytokine Research 1995; 15:1047-1051.
11. Greenberg Pl, Mosny SA. Cytotoxic effects of interferon in vitro on granulocytic progenitor cells. Cancer Res 1977; 37:1794-1799.
12. Grotendorst GR, Smale G, Pencev D. Production of Transforming growth factor beta by human peripheral blood monocytes and neutrophils. J Cell Phys 1989; 140:396-402.
13. Fava RA, Olsen NJ, Postlethwaite AE et al. Transforming Growth factor β1 (TGFβ1) induced neutrophil recruitment to synovial tissues: implications for TGFβ-driven synovial inflammation and hyperplasia. J Exp Med 1993; 173:1121-1132.
14. Zurier RB, Hoffstein S, Weissman G. Cytochalasin B: effects on lysosomal enzyme release in rabbit neutrophils. Proc Natl Acad Sci USA 1973; 70:844-848.
15. Border WA, Noble NA. Transforming growth factor beta in tissue fibrosis. N Engl J Med 1994; 331:1286-1292.
16. Firenstein GS, Zvaifler NJ. Rheumatoid arthritis: disease of disordered immunity. In: Gallin JI, Goldstein IM, Snyderman A, eds. Inflammation: Basic Principles and Clinical Correlates. 2nd edition. New York: Raven Press, 1992:959-975.
17. Brandes ME, Mai UE, Ohura K et al. Type-I transforming growth factor beta receptors on neutrophils mediate chemotaxis to transforming growth factor beta. J Immunol 1991; 147:1600-1606.
18. Parekh, T, Saxena B, Reibman J et al. Neutrophil chemotaxis in response to TGFβ isoforms (TGFβ1, TGFβ2, TGFβ3) is mediated by fibronectin. J Immunol 1994; 152:2456-2466.
19. Wahl SM, Hunt DA, Wakefield L et al. Transforming growth factor beta induces monocyte chemotaxis and growth factor production. Proc Natl Acad Sci USA 1987; 84:5788-5792.
20. Lindemann A, Riedel D, Oster W et al. Granulocyte-Macrophage Colony-Stimulating Factor induces cytokine secretion by human polymorphonuclear leukocytes. J Clin Invest 1989; 83:1308-1312.

21. Collart MA, Belin D, Vassalli JD et al. Gamma interferon enhances macrophage transcription of the tumor necrosis factor/cachectin, interleukin 1, and urokinase genes, which are controlled by short-lived repressors. J Exp Med 1986; 164:2113-2118.
22. Mandi Y, Degre M, Beladi I. Involvement of tumor necrosis factor in human granulocyte-mediated killing of WEHI 164 cells. Int Arch Allergy Appl Immunol 1989; 90:411-413.
23. Dubravec DB, Spriggs DR, Mannick JA et al. Circulating human peripheral blood granulocytes synthesize and secrete tumor necrosis factor alpha. Proc Natl Acad Sci USA 1990; 87:6758-6761.
24. Djeu JY, Serbousek D, Blanchard DK. Release of tumor necrosis factor by human polymorphonuclear leukocytes. Blood 1990; 76:1405-1409.
25. Haziot A, Tsuberi BZ, Goyert SM. Neutrophil CD14: biochemical properties and role in the secretion of tumor necrosis factor-α in response to lipopolysaccharide. J Immunol 1993: 150;5556-5565.
26. Wright SD, Ramos RA, Tobias PS et al. CD14, a receptor for complexes of lipoplysaccharide (LPS) and LPS binding protein. Science 1990; 249:1431-1436.
27. Mandi Y, Endresz V, Krenacs L et al. Tumor necrosis factor production by human granulocytes. Int Arch Allergy Appl Immunol 1991; 96:102-106.
28. Djeu JY. TNF production by neutrophils. In: Beutler B, ed. Tumor Necrosis Factor: The Molecules and Their Emerging Role in Medicine. New York: Raven Press, 1992:531-537.
29. Bazzoni F, Cassatella MA, Laudanna C et al. Phagocytosis of opsonized yeast induces TNFα mRNA accumulation and protein release by human polymorphonuclear leukocytes. J Leuk Biol 1991; 50:223-228.
30. Cassatella MA, Bazzoni F, D'Andrea A et al. Studies on the production of proinflammatory cytokines and on the modulation of gene expression for some NADPH oxidase components by phagocytosing human neutrophils. Fund Clin Immunol 1993; 1:99-106.
31. Palma C, Cassone A, Serbousek D et al. Lactoferrin release and Interleukin-1, Interleukin 6, and Tumor necrosis factor production by human polymorphonuclear cells stimulated by various lipopolysaccharides: relationship to growth inhibition of *Candida albicans*. Infect Immun 1992; 60:4604-4611.
32. Scuderi P, Nez PA, Duerr ML et al. Cathepsin G and leukocyte elastase inactivate human tumor necrosis factor and lymphotoxin. Cell Immunol 1991; 135:299-313.
33. Van kessel KP, Van Strip JA, Verhoef J. Inactivation of recombinant human tumor necrosis factor alpha by proteolytic enzymes released from stimulated human neutrophils. J Immunol 1991; 147:3862-3868.
34. Steadman R, Petersen MM, Topley N et al. Differential augmentation by recombinant human tumor necrosis factor-α of neutrophil responses to particulate zymosan and glucan. J Immunol 1990; 144:2712-2718.
35. Arnold R, Scheffer J, Konig B et al. Effects of *Listeria monocytogenes* and *Yersinia enterocolitica* on cytokine gene expression and release from hu-

man polymorphonuclear granuocytes and epithelial (HEp-2) cells. Infect Immun 1993; 61:2545-2552.
36. Wei S, Blanchard DK, Liu JH et al. Activation of tumor necrosis factor-α production from human neutrophils by IL-2 via IL-2Rβ. J Immunol 1993; 150:1979-1987.
37. Djeu JY, Liu JH, Wei S et al. Function asociated with IL-2 receptor-β expression on human neutrophils: mechanism of activation of antifungal activity against *Candida albicans* by IL-2. J Immunol 1993; 150:960-970.
38. Pericle F, Liu Jh, Diaz JI et al. Interleukin-2 prevention of apoptosis in human neutrophils. Eur J Immunol 1994; 24:440-444.
39. Takeshita T, Asao H, Ohtani K et al. Cloning of the γ chain of the human IL-2 receptor. Science 1992; 257:379-382.
40. Asao H, Takeshita T, Ishii N et al. Reconstitution of functional IL-2 receptor complexes on fibroblastoid cells: involvement of the cytoplasmic domain of the γ chain in two distinct signaling pathways. Proc Natl Acad Sci Usa 1993; 90:4127-4131.
41. Ishii N, Takeshita T, Kimura Y et al. Epression of the IL-2 receptor γ chain on various populations in human peripheral blood. Int Immunol 1994; 6:1273-1277.
42. Philips JH, Takeshita T, Sugamura K et al. Activation of natural killer cells via the p75 interleukin-2 receptor. J Exp Med 1989; 170:291-296.
43. Liu JH, Wei S, Ussery DW et al. Expression of the IL-2 receptor γ chain on human neutrophils. Blood 1994; 84:3780-3875.
44. Girard D, Gosselin J, Heitz D et al. Effects of IL-2 on gene expression in human neutrophils. Blood 1995; 86:1170-1176.
45. Takeichi O, Saito I, Tsurumachi T et al. Human polymorphonuclear leukocytes derived from chronically inflamed tissue express inflammatory cytokines in vivo. Cell Immunol 1995; 156;296-309.
46. Terashima T, Soejima K, Waki Y et al. Neutrophils activated by G-CSF suppress TNFα release from monocytes stimulated by endotoxin. Am J Resp Cell Mol Biol 1995; 13:69-73.
47. Meda L, Gasperini S, Ceska M et al. Modulation of proinflammatory cytokine release from human polymorphonuclear leukocytes by gamma interferon. Cell Immunol 1994; 57:448-461.
48. Wang P, Wu P, Anthes JC et al. Interleukin-10 inhibits Interleukin-8 production in human neutrophils. Blood 1994; 83:2678-2683.
49. Kasama T, Strieter RM, Lukacs NW et al. Interferon gamma modulates the expression of neutrophil-derived chemokines. J Invest Med 1995; 43:58-67.
50. Laudanna C, Constantin G, Baron PL et al. Sulfatides trigger increase of cytosolic free calcium and enhanced expression of tumor necrosis factor-α and interleukin-8 mRNA in human neutrophils. Evidence for a role of L-selectin as a signaling molecule. J Biol Chem 1994; 269:4021-4026.
51. Stoppacciaro A, Melani C, Parenza M et al. Regression of an established tumor genetically modified to release granulocyte colony-stimulating fac-

tor requires granulocyte-T cell cooperation and T cell-produced IFNγ. J Exp Med 1993; 178:151-161.
52. Colombo MP, Lombardi L, Melani C et al. Hypoxic tumor cell death and modulation of endothelial adhesion molecules in the regression of G-CSF transduced tumors. Am J Pathol 1996, in press.
53. Pober JS, Gimbrone MA, Lapierre LA et al. Overlapping patterns of activation of human endothelial cells by interleukin-1, tumor necrosis factor, and immune interferon. J Immunol 1986; 137:1893-1896.
54. Gamble JR, Harlan JM, Klebanoff SJ et al. Stimulation of the adherence of neutrophils to umbilical vein endothelium by human recombinant tumor necrosis factor. Proc Natl Acad Sci USA 1985; 82:8667-8671.
55. Klebanoff SJ, Vadas MA, Harlan JM et al. Stimulation of neutrophils by tumor necrosis factor. J Immunol 1986; 136:4220-4225.
56. Shalaby MR, Aggarwal BB, Rinderknecht E et al. Activation of human polymorphonuclear neutrophil functions by IFNγ and TNFs. J Immunol 1985; 135:2069-2073.
57. Richter J, Andersson T, Olsson I. Effect of TNF and granulocyte-macrophage colony-stimulating factor on neutrophil degranulation. J Immunol 1989; 142:3199-3205.
58. Nathan CF. Neutrophil activation on biological surfaces. J Clin Invest 1987; 80:1550-1560.
59. Della Bianca V, Dusi S, Nadalini KA et al. Role of 55- and 75 kDa TNF receptors in the potentiation of Fc-mediated phagocytosis in human neutrophils. Biochem Biophys Res Commun 1995; 214:44-50.
60. Dinarello CA. The interleukin-1 family: 10 years of discovery. FASEB J 1994; 8:1314-1324.
61. Hanson DF, Murphy PA, Windle BE. Failure of rabbit neutrophils to secrete endogenous pyrogen when stimulated with *Staphylococci*. J Exp Med 1980; 151:1360-1371.
62. Windle BE, Murphy PA, Cooperman S. Rabbit polymorphonuclear leukocytes do not secrete endogenous pyrogens or interleukin 1 when stimulated by endotoxin, polyinosine: polycytosine, or muramyl dipeptide. Infect Immun 1983; 39:1142-1146.
63. Jupin C, Parant M, Chedid L et al. Enhanced oxidative burst without interleukin 1 production by normal human polymorphonuclear leukocytes primed with muramyl dipeptides. Inflammation 1987; 11:53-61.
64. Goto F, Nakamura S, Goto K et al. Production of a lymphocyte proliferating potentiating factor by purified polymorphonuclear leukocytes from mice and rabbits. Immunology 1984; 53:683-692.
65. Tiku K, Tiku ML, Skosey JL. Interleukin 1 production by human polymorphonuclear neutrophils. J Immunol 1986; 136:3677-3685.
66. Lindemann A, Riedel D, Oster W et al. Granulocyte-Macrophage Colony-Stimulating Factor induces Interleukin-1 production by humam polymorphonuclear neutrophils. J Immunol 1988; 140:837-839.
67. Dularay B, Westacott CI, Elson CJ. IL-1 secreting cell assay and its ap-

plication to cells from patients with rheumatoid arthritis. Br J Rheum 1992; 31:19-24.
68. Goh K, Furusawa S, Kawa Y et al. Production of Interleukin-1-alpha and -beta by human peripheral polymorphonuclear neutrophils. Int Arch Allergy Appl Immunol 1989; 88:297-303.
69. Lord PCW, Wilmoth LMG, Mizel SB et al. Expression of interleukin-1 alpha and β genes by human blood polymorphonuclear leukocytes. J Clin Invest 1991; 87:1312-1321.
70. Jack RM, Fearon DT. Selective synthesis of mRNA and proteins by human blood neutrophils. J Immunol 1988; 140:4286-4293.
71. Bainton DF, Ullyot JL, Farquhar MG. The development of neutrophilic polymorphonuclear leukocytes in human bone marrow. J Exp Med 1971; 134:907-934.
72. McCall CE, Grosso-Wilmoth LM, LaRue K et al. Tolerance to endotoxin-induced expression of the Interleukin-1β gene in blood neutrophils of humans with the sepsis syndrome. J Clin Invest 1993; 91:853-861.
73. Fasano MB, Cousart S, Neal S et al. Increased expression of the interleukin-1 receptor on blood neutrophils of humans with the sepsis syndrome. J Clin Invest 1991; 88:1452-1459.
74. Hsi ED, Remick DG. Monocytes are the major producers of Interleukin-1b in an ex vivo model of local cytokine production. J Interferon and Cytokine Research 1995; 15:89-94.
75. Malyak M, Smith MF, Abel AA et al. Peripheral blood neutrophil production of IL-1ra and IL-1β. J Clin Immunol 1994; 14:20-30.
76. Marucha PT, Zeff RA, Kreutzer DL. Cytokine regulation of IL-1β gene expression in the human polymorphonuclear leukocyte. J Immunol 1990; 145:2932-2937.
77. Contrino J, Krause PJ, Slover N et al. Elevated interleukin 1 expression in human neonatal neutrophils. Pediatr Res 1993; 34:249-252.
78. Walters JD, Cario AC, Leblebicioglu B et al. An inhibitor of polyamine biosynthesis impairs human polymorphonuclear leukocyte priming by TNFα. J Leuk Biol 1995; 57:282-286.
79. Cassone A, Palma C, Djeu JY et al. Anticandidal activity and interleukin-1β and interleukin 6 production by polymorphonuclear leukocytes are preserved in subjects with AIDS. J Clin Microbiol 1993; 31:1354-1357.
80. Beaulieu AD, Paquin R, Gosselin J. Epstein-Barr virus modulates de novo protein synthesis in human neutrophils. Blood 1995; 86:2789-2798.
81. Roberge CJ, Grassi J, de Medicis R et al. Crystal-neutrophil interactions lead to IL-1 synthesis. Agents Actions 1991; 34:38-41.
82. Roberge CJ, de Medicis R, Dayer JM et al. Crystal-induced neutrophil activation. V. Differential production of biologically active IL-1 and IL-1 receptor antagonist. J Immunol 1994; 152:5485-5494.
83. Dinarello CA. ELISA kits based on monoclonal antibodies do not measure total IL-1β synthesis. J Immunol Methods 1992; 148:255-259.
84. Tiku K, Tiku ML, Liu S et al. Normal human neutrophils are a source of a specific interleukin 1 inhibitor. J Immunol 1986; 136:3686-3692.
85. Eisenberg SP, Evans RJ, Arend WP et al. Primary structure and func-

tional expression from complementary DNA of a human interleukin 1 receptor antagonist. Nature 1990; 343:341-343.
86. Carter DB, Deibel MR, Dunn CJ et al. Purification, cloning, expression and biological of an IL-1 receptor antagonist characterization of an IL-1 receptor anatgonist protein. Nature 1990; 344:633-638.
87. McColl SR, Paquin R, Ménard C et al. Human neutrophils produce high levels of the interleukin 1 receptor antagonist in response to Granulocyte/Macrophage Colony stimulating factor and tumor necrosis factor. J Exp Med 1992; 176:593-598.
88. McColl SR, Paquin R, Beaulieu AD. Selective synthesis and secretion of a 23 KD protein by neutrophils following stimulation with GM-CSF and TNFα. Biochem Byophis Res Commun 1990; 172:1209-1216.
89. Beaulieu AD, Paquin R, Rathanaswami P et al. Nuclear signaling in human neutrophils. Stimulation of RNA synthesis in response to a limited number of proinflammatory agonists. J Biol Chem 1992; 267:426-432.
90. Serra MC, Calzetti F, Ceska M et al. Effect of substance P on superoxide anion and IL-8 production by human PMN. Immunology 1994; 82:63-69.
91. Wright SD. CD14 and innate recognition of bacteria. J Immunol 1995; 155:6-8.
92. Ulich TR, Guo K, Yin S et al. Endotoxin-induced cytokine gene expression in vivo. IV. Expression of interleukin 1α/β and interleukin 1 receptor antagonist mRNA during endotoxemia and during endotoxin-initiated local acute inflammatiom. Am J Pathol 1992; 141:61-68.
93. Re F, Mengozzi M, Muzio M et al. Expression of interleukin 1 receptor antagonist by human circulating polymorphonuclear cells. Eur J Immunol 1993; 23:570-573.
94. Cassatella MA, Meda L, Gasperini S et al. Interleukin 10 up-regulates IL-1 receptor antagonist production from lipopolysaccharide-stimulated human polymorphonuclear leukocytes by delaying mRNA degradation. J Exp Med 1994; 179:1695-1699.
95. Poutsiaka DD, Clark BD, Vannier E et al. Production of IL-1 receptor antagonist and IL-1β by peripheral blood mononuclear cells is differentially regulated. Blood 1991; 78:1275-1281.
96. Arend PW, Smith Jr.MF, Janson RW et al. IL-1 receptor antagonist and IL-1β production in human monocytes are regulated differently. J Immunol 1991; 147:1530-1536.
97. Chang DM. Cellular signals for the induction of human Interleukin-1 receptor anatgonist. Clin Immunol Immunopatol 1995; 74:23-30.
98. Muzio M, Re F, Sironi M et al. Interleukin-13 induces the production of Interleukin-1 receptor antagonist (IL-1ra) and the expression of the mRNA for the intracellular (keratinocyte) form of IL-1ra in human myelomonocytic cells. Blood 1994; 83:1738-1743.
99. Muzio M, Sironi M, Polentarutti N et al. Induction by transforming growth factor-β1 of the interleukin-1 receptor antagonist and of its intracellular form in human polymorphonuclear cells. Eur J Immunol 1994; 24:3194-3198.
100. Muzio M, Polentarutti N, Sironi M et al. Cloning and characterization

of a new isoform of the interleukin-1 receptor antagonist. J Exp Med 1995; 182:623-628.
101. Marie C, Pitton C, Fitting C et al. IL-10 and IL-4 synergize with TNFα to induce IL-1ra production by human neutrophils. Cytokine 1996; in press.
102. Arend WP. Interleukin-1 receptor antagonist. Adv Immunol 1993; 54:167-227.
103. Cicco NA, Lindemann A, Content J et al. Inducible production of interleukin-6 by human neutrophils: role of Granulocyte-Macrophage Colony-Stimulating Factor and tumor necrosis factor alpha. Blood 1990; 75:2049-2052.
104. Kato K, Yokoi T, Takano N et al. Detection by in situ hybridization and phenotypic characterization of cell expressing IL-6 mRNA in human stimulated blood. J Immunol 1990; 144:1317-1322.
105. Melani C, Mattia GF, Silvani A et al. Interleukin-6 expression in human neutrophil and eosinophil peripheral blood granulocytes. Blood 1993; 81:2744-2749.
106. Arnold R, Werner F, Humbert B et al. Effect of respiratory syncytial virus-antibody complexes on cytokine (IL-8, IL-6, TNFα) release and respiratory burst in human granulocytes. Immunology 1994; 82:184-191.
107. Wertheim WA, Kunkel SL, Standiford TJ et al. Regulation of neutrophil-derived IL-8: the role of prostaglandin E_2, dexamethasone, and IL-4. J Immunol 1993; 151:2166-2175.
108. Strieter RM, Kasahara K, Allen R et al. Human neutrophils exhibit disparate chemotactic gene expression. Biochem Biophys Res Commun 1990; 173:725-730.
109. Bazzoni F, Cassatella MA, Rossi F et al. Phagocytosing neutrophils produce and release high amounts of the neutrophil activating peptide 1/Interleukin 8. J Exp Med 1993; 173:771-774.
110. Pakianathan DR. Extracellular matrix proteins and leukocytes. J Leuk Biol 1995; 57:699-702.
111. Cassatella MA, Bazzoni F, Ceska M et al. Interleukin 8 production by human polymorphonuclear leukocytes. The chemoattractant formyl-Methionyl-Leucyl-Phenylalanine induces the gene expression and release of interleukin 8 through a pertussis toxin sensitive pathway. J Immunol 1992; 148:3216-3220.
112. Ayesh SK, Azar Y, Babior BM et al. Inactivation of interleukin-8 by the C5a-inactivating protease from serosal fluid. Blood 1993; 81:1424-1427.
113. Ember JA, Sanderson SD, Hugli TE et al. Induction of IL-8 synthesis from monocytes by human C5a anaphylotoxin. Am J Pathol 1994; 144:393-403.
114. McCain RW, Holden EP, Blackwell TR et al. Leukotriene B_4 stimulates human polymorphonuclear leukocytes to synthesize and release Interleukin-8 in vitro. Am J Respir Cell Mol Biol 1994; 10:651-657.
115. Strieter RM, Kasahara K, Allen RM et al. Cytokine-induced neutrophil-derived Interleukin-8. Am J Pathol 992; 141:397-407.
116. Au B, Williams TJ, Collins PD. Zymosam-induced IL-8 release from

human neutrophils involves activation via the CD11b/CD18 receptor and endogenous platelet-activating factor as an autocrine modulator. J Immunol 1994; 152:5411-5419.
117. Baggiolini M, Dewald B, Moser B. Interleukin-8 and related chemotactic cytokines-CXC and CC chemokines. Adv Immunol 1994; 55:97-179.
118. Konig W, Schonfeld W, Raulf M et al. The neutrophil and leukotrienes-role in health and disease. Eicosanoids 1990; 3:1-22.
119. Fujishima S, Hoffman AR, Vu T et al. Regulation of neutrophil interleukin 8 gene expression and protein secretion by LPS, TNFα and IL-1β. J Cell Physiol 1993; 154:478-485.
120. Cassatella MA, Meda L, Bonora S et al. Interleukin 10 inhibits the release of proinflammatory cytokines from human polymorphonuclear leukocytes. Evidence for an autocrine role of TNF and IL-1β in mediating the production of IL-8 triggered by lipopolysaccharide. J Exp Med 1993; 178:2207-2211.
121. McCain RW, Dessypris EN, and Christman JW. GM-CSF stimulates human polymorphonuclear leukocytes to produce Interleukin-8 in vitro. Am J Respir Cell Mol Biol 1993; 8:28-34.
122. Takahashi GW, Andrews DF, Lilly MB et al. Effect of GM-CSF and IL-3 on IL-8 production by human neutrophils and monocytes. Blood 1993; 81:357-364.
123. Cavaillon JM, Marie C, Pitton C et al. Regulation of neutrophil derived IL-8 production by anti-inflammatory cytokines (IL-4, IL-10 and TGFb). In: Faist E, ed. Proceedings 3rd International Congress on the Immune Consequences of Trauma, Shock and Sepsis. Munich, Pabst Science Publishers, 1996, in press.
124. Wei S, Liu JH, Blanchard DK et al. Induction of IL-8 gene expression in human polymorphonuclear neutrophils by recombinant IL-2. J Immunol 1994; 152:3630-3636.
125. Cassatella MA, Guasparri I, Ceska M et al. Interferon-γ inhibits interleukin-8 production by human polymorphonuclear leukocytes. Immunology 1993; 78:177-184.
126. Ross JI, Cain JA, and Lachmann PJ. Membrane complement receptor type three (CR3) has lectin-like properties analogous to bovin conglutinin and functions as a receptor for zymosan and rabbit erythrocytes as well as a receptor iC3b. J Immunol 1985; 134:3307-3315.
127. Tool AT, Verhoeven AJ, Roos D et al. Platelet-activating factor (PAF) acts as an intercellular messenger in the changes of cytosolic free Ca^{++} in human neutrophils induced by opsonized particles. FEBS lett 1989; 259:209-212.
128. Nieto ML, Velasco S, Crespo MS. Biosynthesis of PAF in human polymorphonuclear leukocytes. Involvement of cholinephosphotransferase pathway in response to phorbol esters. J Biol Chem 1988; 263:2217-2222.
129. Kuhns D, Gallin JI. Increased cell-associated IL-8 in human exudative and 53187-treated peripheral blood neutrophils. J Immunol 1995; 154:6556-6562.

130. Cassatella MA, Aste M, Calzetti F et al. Studies on the regulatory mechanisms of Interleukin-8 gene expression in resting and IFNγ-treated neutrophils. Evidence on the capability of staurosporine of inducing the production of IL-8 by human neutrophils. Biochem Biophys Res Commun 1993; 190:660-667 (published erratum in 1993; 192:324).
131. Borregaard N, Christensen O, Bjerrum OW et al. Identification of a highly mobilizable subset of human neutrophil intracellular vesicles that contains tetranectin and latent alkaline phosphatase. J Clin Invest 1990; 85:408-416.
132. Jorgensen C, Sany J. Modulation of the immune response by the neuroendocrine axis in rheumatoid arthritis. Clin Exp Rheumatol 1994; 12:435-441.
133. Serra MC, Bazzoni F, Della Bianca V et al. Activation of human neutrophils by substance P: effect on oxidative metabolism, exocytosis, cytosolic Ca^{2+} concentration and inositol phosphate formation. J Immunol 1988; 141:2118-2124.
134. Wozniak A, McLennan G, Betts WH et al. Activation of human neutrophils by substance P. Effect on fMLP-stimulated oxidative and arachidonic acid metabolism and on antibody-dependent cell-mediated cytotoxicity. J Immunol 1989; 68:359-367.
135. Ruegg UT, Burgess GM. Staurosporine, K-252 and UCN-O1: potent but nonspecific inhibitors of protein kinase. Trends Pharm Sci 1989; 10:218-220.
136. Konig B, Koller M, Prevost G et al. Activation of human effector cells by different bacterial toxins (leukocidin, alveolysin, and erythrogenic toxin A): generation of interleukin-8. Infect Immun 1994; 62:4831-4837.
137. Inoue H, Massion P, Ueki IF et al. *Pseudomonas* stimulates Interleukin-8 mRNA expression selectively in airway epithelium, in gland ducts, and in recruited neutrophils. Am J Respir Cell Mol Biol 1994; 11:651-663.
138. Konig B, Ceska M, Konig W, et al. Effect of *Pseudomonas aeruginosa* on interlerukin-8 release from human phagocytes. Int Arch Allergy Immunol 1995; 106:357-365.
139. Hachicha M, Naccache PH, McColl SR. Inflammatory microcrystal differentially regulate the secretion of macrophage inflammatory protein 1 and interleukin 8 by human neutrophils: a possible mechanism of neutrophil recruitment to sites of inflammation in synovitis. J Exp Med 1995; 182:2019-2025.
140. Kasahara K, Strieter RM, Chensue SW et al. Mononuclear cell adherence induces neutrophil chemotactic factor/interleukin-8 gene expression. J Leuk Biol 1991; 50:287-295.
141. Cassatella MA, Meda L, Gasperini S et al. Interleukin-12 production by human polymorphonuclear leukocytes. Eur J Immunol 1995; 25:1-5.
142. Ma X, Chow JM, Gri G et al. The IL-12 p40 gene promoter is primed by IFNγ in monocytic cells. J Exp Med 1996; 183:147-157.
143. D'Andrea A, Aste Amezaga M, Valiante NM et al. Interleukin-10 inhibits human lymphocyte IFN-γ production by suppressing natural killer cell stimulatory factor/Interleukin-12 synthesis in accessory cells. J Exp Med

1993; 178:1041-1048.
144. D'Andrea A, Rengaraju M, Valiante NM et al. Production of natural killer cell stimulatory factor (interleukin 12) by peripheral blood mononuclear cells. J Exp Med 1992; 176:1387-1398.
145. Hayes MP, Wang J, Norcross MA. Regulation of Interleukin-12 expression in human monocytes: selective priming by interferon-γ of lipopolysaccharide-inducible p35 and p40 genes. Blood 1995; 86:646-650.
146. Nathan CF. Interferon and inflammation. In: Gallin JI, Goldstein IM, Snyderman R, eds. Inflammation: Basic Principles and Clinical Correlates. 2nd edition. New York: Raven Press, 1992:265-290.
147. Haskill S, Peace A, Morris J et al. Identification of three related human GRO genes encoding cytokine functions. Proc Natl Acad Sci USA 1990; 87:7732-7736.
148. Iida N, Grotendorst GR. Cloning and sequencing of a new *gro* transcript from activated human monocytes: expression in leukocytes and wound tissue. Mol Cell Biol 1990; 10:5596-5599.
149. Gasperini S, Calzetti, F, Russo MP, De Gironcoli M, Cassatella MA. Regulation of GROα production in human granulocytes. J Inflamm 1995; 45:143-151.
150. Bordoni R, Fine R, Murray D et al. Characterization of the role of melanoma growth factor stimulatory activity (MGSA) in the growth of normal melanocytes, nevocytes, and malignant melanocytes. J Cell Biochem 1990; 44:207-219.
151. Koch AE, Kunkel SL, Shah MR et al. Growth-related gene product α. A chemotatcic cytokine for neutrophils in rheumatoid arthritis. J Clin Invest 1995; 155:3660-3666.
152. Derynck R, Balentien E, Han JH et al. Recombinant expression, biochemical characterization, and biological activities of the human MGSA/gro protein (published erratum appears in Biochemistry 1991 Jan 15:594). Biochemistry 1990; 29:10225-10233.
153. Schroder JM, Persoon NL, Christophers E. Lipopolysaccharide-stimulated human monocytes secrete, apart from neutrophil-activating peptide 1/interleukin-8, a second neutrophil-activating protein: NH2-terminal amino acid sequence identity with melanoma growth stimulatory activity. J Exp Med 1990; 171:1091-1100.
154. Moser B, Clark-Lewis I, Baggiolini M. Neutrophil-activating properties of the melanoma growth-stimulatory activity. J Exp Med 1990; 171:1797-1802.
155. Huang S, Paulauskis JD, Godleski JJ et al. Expression of macrophage inflammatory protein-2 and KC mRNA in pulmonary inflammation. Am J Pathol 1992; 41:981-988.
156. Kasama T, Strieter RM, Standiford TJ et al. Expression and regulation of human neutrophil-derived macrophage inflammatory protein 1-alpha. J Exp Med 1993; 178:63-72.
157. Kasama T, Strieter RM, Lukacs NW et al. Regulation of neutrophil-derived chemokine expression by IL-10. J Immunol 1994; 152:3559-3569.

158. Van Damme J, Proost P, Put W et al. Induction of monocyte chemotactic proteins MCP-1 and MCP-2 in human fibroblasts and leukocytes by cytokines and cytokine inducers. Chemical synthesis of MCP-2 and development of a specific RIA. J Immunol 1994; 152:5495-5502.
159. Burn TC, Petrovick MS, Hohaus S et al. Monocyte chemoattractant protein-1 gene is expressed in activated neutrophils and retinoic acid-induced human myeloid cell lines. Blood 1994; 84:2776-2783.
160. Sakanashi Y, Takeya M, Yoshimura T et al. Kinetics of macrophage subpopulations and expression of monocyte chemoattractant protein-1 (MCP-1) in bleomycin-induced lung injury of rats studied by a novel monoclonal antibody against rat MCP-1. J Leuk Biol 1994; 56:741-750.
161. D'Angio CT, Sinkin RA, LoMonaco MB et al. Interleukin-8 and monocyte chemoattractant protein-1 mRNAs in oxygen-injured rabbit lung. Am J Physiol 1995; 12:L826-L831.
161a. Taub DD, Lloyd AR, Conlon K et al. Recombinant human interferon-inducible protein 10 is a chemoattractant for human monocytes and T lymphocytes and promotes T cell adhesion to endothelial cells. J Exp Med 1993; 177:1809-1814.
162. Smith C, Davis T, Anderson D et al. A receptor for tumor necrosis factor defines an unusual family of cellular and viral proteins. Science 1990; 248:1019-1023.
163. Del Prete G, De Carli M, Almerigogna F et al. Preferential expression of CD30 by human CD4+ T cell producing Th2-type cytokines. FASEB J 1995; 9:81-86.
164. Manetti R, Annunziato F, Biagiotti R et al. CD30 expression by CD8+ T cell producing type 2 helper cytokine. Evidence for large numbers of CD8+ CD30+ T cell clones in human immunodeficiency virus infection. J Exp Med 1994; 180:2407-2412.
165. Smith CA, Gruss H-J, Davis T et al. CD30 antigen, a marker for Hodgkin's lymphoma, is a receptor whose ligand defines an emerging family of cytokine with homology to TNF. Cell 1993; 73:1349-1360.
166. Smith CA, Farrah T, Goodwin RG. The TNF receptor superfamily of cellular and viral proteins: activation, costimulation and death. Cell 1994; 76:959-962.
167. Younes A, Jaing S, Gruss H et al. Expression of the CD30 ligand (CD30L) on human peripheral blood lymphocytes from normal donors and patients with lymphoid malignancies. Blood 1994; 84(Suppl.1):628 (Abstr.).
168. Gruss H-J, Boiani N, Williams DE et al. Pleiotropic effects of the CD30 ligand on CD30-expressing cells and lymphoma cell lines. Blood 1994; 83:2045-2051.
169. McDonald PP, Cassatella MA, Bald A et al. CD30 ligation induces nuclear factor-kB activation in human T cell lines. Eur J Immunol 1995; 25:2870-2876.
170. Del Prete G, De Carli M, D'Elios M et al. CD30-mediated signaling

promotes the development of human T helper type 2-like T cells. J Exp Med 1995; 182:1655-1661.
171. Gruss HJ, DaSilva N, Hu ZB et al. Expression and regulation of CD30 ligand and CD30 in human leukemia-lymphoma cell lines. Leukemia 1994; 8:2083-2094.
172. Biswas P, Smith CA, Goletti D et al. Cross-linking of CD30 induces HIV expression in chronically infected T cells. Immunity 1995; 2:587-596.
173. Ho JL, He S, Hu A et al. Neutrophils from human immunodeficiency virus (HIV)-seronegative donors induce HIV replication from HIV-infected patients mononuclear cells and cell lines: an in vitro model of transmission facilitated by *Chlamydia trachomatis*. J Exp Med 1995; 181:1493-1505.
174. Kita H, Ohnishi T, Okubo Y et al. Granulocyte/Macrophage Colony-stimulating Factor and Interleukin 3 release from human peripheral blood eosinophils and neutrophils. J Exp Med 1991; 174:745-748.
175. Ramenghi U, Ruggieri L, Dianzani I et al. Human peripheral blood granulocytes and myeloid leukemic cell lines express both transcripts encoding for stem cell factor. Stem Cells 1994; 12:521-526.
176. Galli SJ, Zsebo KM, Geissler EN. The Kit ligand, stem cell factor. Adv Immunol 1994; 55:1-96.
177. Bry K, Hallmann M, Lappalainen U. Cytokines released by granulocytes and mononuclear cells stimulate amnion cell prostaglandin E_2 production. Prostaglandins 1994; 48:389-399.
178. Bortolami M, Carlotto C, Fregona I, et al. Effects of interferon-α on the production of tumor necrosis factor-α by polymorphonuclear cells. Fund Clin Immunol 1995; 3:153-156.
179. Jenkins JK, Malyak M, and Arend, WP. The effects of Interleukin-10 on Interleukin-1 receptor antagonist and Interleukin-1β production in human monocytes and neutrophils. Lymphokine and Cytokine Research 1994; 13:47-54.
180. Marcolongo R, Zambello R, Trentin L et al. Childhood onset cyclic neutropenia: G-CSF therapy restores neutrophil count but does not influence superoxide anion and cytokine by neutrophils. Br J Hematol 1995; 89:277-281.

CHAPTER 5

MODULATION OF CYTOKINE PRODUCTION IN HUMAN NEUTROPHILS

In addition to the growing number of stimuli that are able to elicit cytokine expression in neutrophils in vitro, another important aspect emerging from the literature is the identification of mediators, or of various types of interactions, that can modulate cytokine production by neutrophils. An increasing number of studies have in fact documented that cytokine expression by polymorphonuclear leukocytes (PMN) may depend also on the "functional" state of the cell at the moment the stimulus is received. This functional state, in turn, is influenced by inflammatory or immunomodulating agents, that are known to affect other neutrophil effector responses.[1] The mechanisms whereby these agents potentiate or down-regulate cytokine production is (in most cases) still a mystery, the elucidation of which being rendered even more difficult by the fact that these agents, usually, but not always, do not directly trigger neutrophil functional responses. The majority of the studies addressing control of cytokine production by PMN have been conducted using lipopolysaccharide (LPS) as a stimulus, probably because of its being a potent inducer of the production of many different cytokines. However, other important biological agonists for neutrophils, such as Tumor Necrosis Factor-α (TNFα), formyl-methionyl-leucyl-phenylalanine (fMLP) and IgG-opsonized *S.cerevisiae* (Y-IgG) have also been used as stimuli. Table 3.2 summarizes very schematically the modulatory effects of Interferon-γ (IFNγ) and Interleukin-10 (IL-10) on the production of cytokines by neutrophils, mainly based on work done in my laboratory. Although, IFNγ and IL-10 are the two molecules whose effects have been mostly investigated until now, the effects of IL-4 and of other substances have also been studied.

Cytokines Produced by Polymorphonuclear Neutrophils: Molecular and Biological Aspects, by Marco A. Cassatella. © 1996 R.G. Landes Company.

EFFECTS OF INTERLEUKIN-4 (IL-4)

As outlined in chapter 2, IL-4 modulates various monocyte functions related to inflammation,[2,3] but is also known to influence specific neutrophil functions.[4] Furthermore, IL-4 is also one of the cytokines which has been shown to moderately modulate the release of specific cytokines from PMN.

Wertheim and colleagues[5] were the first to observe that IL-4 inhibits IL-8 production by PMN. In this regard, an inhibition of about 75% was obtained after 24 h of culture, using 10 ng/ml of IL-4, which also occurred in the presence of suboptimal stimulatory doses of LPS (10 ng/ml). Northern blot analysis demonstrated that treatment of PMN with IL-4 for 2 h resulted in a 52% reduction in steady-state mRNA levels stimulated by LPS. These findings were substantially confirmed by Wang et al,[6] who showed that 10 ng/ml IL-4 inhibited (by approximately 35%) the production of IL-8 from PMN stimulated with LPS (80 ng/ml) for 14 h. In this negative action, IL-4 was shown to be more potent than equivalent concentrations of Transforming Growth Factor-β (TGFβ), but less potent than IL-10.[6] Studies by Cavaillon et al[7] and from our laboratory (Fig. 5.1) also found that IL-4 is less potent than IL-10 in inhibiting the release of IL-8 from stimulated PMN. However, while IL-4 was a poor inhibitor of PMN-derived IL-8 release also in response to Y-IgG, it potentiated the negative effect of IL-10 on the release of cytokines by PMN, including IL-8 (our unpublished observations, and Fig. 5.1). Furthermore, Cavaillon et al[7] showed that IL-4 did not inhibit the release of IL-8 triggered by high doses of TNFα (20 ng/nl); however, in contrast with the data of Wang et al[6] and our own (unpublished observations), they found TGFβ to be not only ineffective in suppressing LPS- or TNFα-induced IL-8 production, but to be able by itself to induce the expression of IL-8 mRNA and protein in neutrophils.[7]

The effects of IL-4 towards IL-1β and IL-1ra mRNA and protein production induced by LPS and other stimuli have been widely investigated by Malyak et al.[8] These authors found that IL-4, in a dose-dependent manner, markedly decreased (by up to 50%) the LPS-induced total IL-1β protein synthesis over a 22 h period. This effect of IL-4 occurred without reducing the LPS-stimulated IL-1β mRNA levels, suggesting that it acted at the post-transcriptional level.[8] Our unpublished experiments confirm that IL-4 indeed suppresses the production of IL-1β from PMN stimulated with LPS (but not with Y-IgG), without reducing the LPS-dependent increase of IL-1β mRNA levels (Fig. 5.2). Moreover, Malyak et al[8] showed that IL-4 alone minimally increased the total synthesis of IL-1ra by PMN, in disagreement with other studies,[9,10] which reported a substantial production of IL-1ra in response to IL-4. However, in the same study, Malyak et al[8] showed that IL-4 in combination with LPS produced a more than additive increase in IL-1ra production, which was paralleled by a relative in-

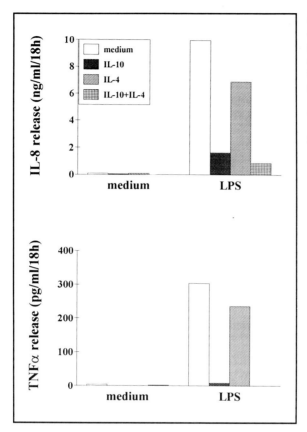

Fig. 5.1. Effect of IL-4 on IL-8 and TNFα extracellular production in LPS-treated neutrophils. PMN were cultured for 18 h in the presence or absence of 100 U/ml IL-4 and/or LPS (1 μg/ml). Cytokines were then measured in the cell-free supernatants.

crease in the steady-state levels of IL-1ra mRNA. Regarding this last effect, we (unpublished observations), as well as Marie et al,[10] obtained identical results. Interestingly, according to the data presented by Marie et al,[10] the production of IL-1ra induced by TNFα was also amplified by IL-4 and IL-10, but not by IL-13. No reports on the effects of IL-4 on the production of TNFα by PMN have been published so far, but our preliminary data suggest that LPS-induced TNFα mRNA expression and release are inhibited by IL-4, although less efficiently than by IL-10 (unpublished results, and Figs. 5.1 and 5.2).

Colotta and coworkers[11] have made important observations on the relationship between neutrophils and IL-4, which I believe are well worth underlining in this context. They showed that IL-4 almost completely abolishes[11] the beneficial effect of exogenous IL-1β on the survival of PMN in culture.[12] Such an effect was due to the fact that IL-4, unlike IL-10 (A. Mantovani, personal communication), augments membrane expression and release of the type II IL-1 receptor (IL-1RII).[11] This receptor is the one predominantly expressed in PMN, and, contrast

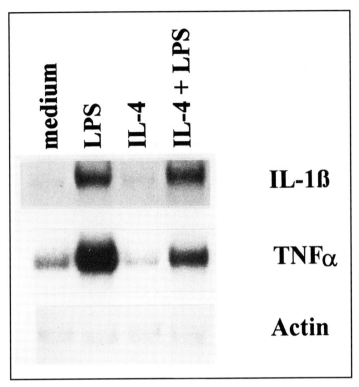

Fig. 5.2. Effect of IL-4 on IL-1β and TNFα mRNA expression in LPS-treated neutrophils. PMN were cultured for 2.5 h, and then subjected to Northern blot analysis.

to the IL-1RI, has no known signaling function.[13] The IL-1-induced survival of PMN was blocked by anti-IL-1RI antibodies.[11] The inhibitory activity of IL-4 was totally abrogated by the presence of blocking antibodies directed against IL-1RII, consistent with a model in which the IL-1RII is an inhibitor of IL-1 function. Released IL-1RII seems to act as a IL-1 decoy target,[11] so that by augmenting the synthesis and shedding of IL-1RII from PMN, IL-4 impedes the positive effects of IL-1β towards neutrophils.[14] Interestingly, soluble IL-1RII is shed within minutes after treatment of PMN and monocytes with chemotactic stimuli and oxygen radicals, indicating that release of this receptor represents an aspect of the complex reprogramming of myelomonocytic cells in response to these mediators.[15] Furthermore, the findings that glucocorticoid and Th2-derived cytokines (IL-4 and IL-13) up-regulate IL-1RII expression and release, is in keeping with the concept that the IL-1RII may represent a physiological pathway of IL-1 inhibition, contributing to the anti-inflammatory properties of Th2-derived cytokines and glucorticoids.[14] In the study of Colotta et al,[11] prolongation of PMN survival induced by LPS[12] was also partially inhibited by IL-4, possibly because of an IL-1RII-mediated blocking effect towards endogenously produced IL-1β in cells stimulated by endotoxin. Whether the same circuit contributes to the suppression by IL-4 on the LPS-induced production of IL-8, is not yet known. Similarly, whether IL-4 down-regulates the LPS-induced IL-8 release because of its inhibition on TNFα production, as demonstrated in the case of IL-10 (see below), remains to be investigated.

EFFECTS OF INTERLEUKIN-10 (IL-10)

Several groups have demonstrated that IL-10 is a very potent modulator of cytokine release from neutrophils. IL-10 is mainly synthesized by monocytes, B cells and certain populations of CD4+ and CD8+ T cells, and as a general rule, has the capacity to attenuate a wide range of inflammatory and immune responses.[16] Very recently, IL-10 was also demonstrated to be also a potent inhibitor of both eosinophil survival and secretion of IL-8, Granulocyte-Macrophage Colony-Stimulating Factor (GM-CSF), and TNFα induced by LPS, but not of eosinophil survival mediated by GM-CSF.[17]

At the end of 1993, we reported that in neutrophils, IL-10 dose-dependently inhibited TNFα, IL-1β and IL-8 secretion triggered by 1 μg/ml LPS and Y-IgG[18] (Fig. 5.3). We found that, in response to LPS, IL-10 almost completely abrogated the release of TNFα and IL-1β at all time points (from 2 to 24 h), but did not significantly influence the production of IL-8 after 2-3 h of neutrophil stimulation. Conversely, IL-10 markedly reduced the extracellular accumulation of IL-8 at 6 h (by 44%) and at 18 h (by 69%).[18] Northern blot analysis revealed that IL-10 also diminished the LPS-induced accumulation of TNFα, IL-1β and IL-8 mRNAs, but only at time points later than 4-5 h[18] (see also Fig. 5.4, which shows kinetics of mRNA expression of several cytokines in PMN). Similar observations have been made by Cavaillon et al[7] in neutrophils, by Donnelly et al[19] in human monocytes, and by Bogdan et al[20] in murine macrophages.

Since it was already known that both TNFα and IL-1β could independently induce IL-8 secretion in PMN,[21,22] we further showed that their combined use produced at least an additive, if not a synergistic, effect on the release of IL-8 from PMN.[18] Importantly, the combined action of TNFα and IL-1β was not only observed with the same doses of TNFα and IL-1β detected in LPS-stimulated PMN supernatants, but it was not inhibited by IL-10 at all.[18] The latter observation has been confirmed recently by Cavaillon et al.[7] On the basis of these findings, and reassured by the success of an identical approach made with whole blood,[23] we examined the possibility that the inhibitory effect of IL-10 on IL-8 release was the result of the suppression of TNFα and IL-1β secretion.[18] By using anti-TNFα and anti-IL-1β neu-

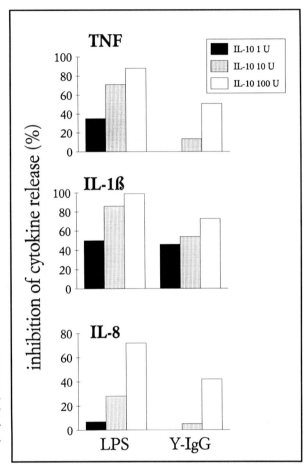

Fig. 5.3. Dose-dependent inhibitory effect of IL-10 on TNFα, IL-1β and IL-8 extracellular release by neutrophils stimulated for 6 h with LPS or IgG-opsonized S.cerevisiae (Y-IgG).

tralizing antibodies, we showed that within the first 6 h of LPS stimulation, anti-TNFα and anti-IL-1β did not significantly influence the extracellular accumulation of IL-8 stimulated by LPS, but that over the remainder of the time course, the release of IL-8 was significantly reduced to levels comparable to, but still slightly lower than, those observed in IL-10-treated cells.[18] Therefore, the experiments performed with IL-10 revealed that the kinetics of IL-8 release in response to LPS consists of two distinct phases: an early one, directly induced by LPS, followed by a late, prolonged phase that appeared to be in part due to the endogenous release of TNFα and IL-1β[18] (see the schemes depicted in Fig. 3.7). The first wave accounts for approximately 20-30% of the total IL-8 secreted by PMN, lasts a few hours, and is inhibitable by IFNγ;[24] the second one is more sustained, leads to an elevated production of IL-8, and is inhibitable by IL-10.[18] Consistent with the relationship between endogenous production of TNFα and IL-1β, and

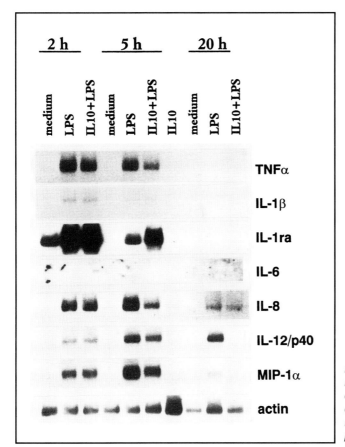

Fig. 5.4. Effect of IL-10 on the mRNA expression for various cytokines in LPS-treated neutrophils.

their capacity to induce the release of IL-8, were our previous results obtained using fMLP as a stimulus of cytokine production.[24a] In response to this chemotactic factor, PMN release neither TNFα nor IL-1β, but only IL-8, which is actively secreted for an initial 2-3 h, after which it rapidly decreases to basal levels. The lack of autocrine/paracrine effects of TNFα or IL-1β, could be one of the reasons explaining the characteristically transient kinetics of fMLP-induced IL-8 release. Furthermore, previous experiments in which the time course of the release of IL-8 by LPS-stimulated whole blood was examined, also identified a biphasic pattern of IL-8 production in that system, and attributed to endogenous TNFα and IL-1 an autocrine role.[23] The sequential production of IL-8 by activated PMN, regulated by TNFα, IL-1β and IL-8 by activated PMN might be finalized to amplify the recruitment and activation of neutrophils during the inflammatory response to LPS. As it will be discussed in chapter 9, the same cytokine-network does not apply in the case of Y-IgG-stimulated PMN.[18]

In a successive work exploring the effects of IL-10 on PMN, we could demonstrate that IL-10 markedly potentiated the extracellular yield of IL-1ra from LPS-stimulated neutrophils (from 2- to 3-fold after 18 h), but not from Y-IgG-treated cells.[25] A possible explanation for this action was put forward, that is, the prolongation of IL-1ra mRNA half-life by IL-10 in LPS-treated PMN.[25] Whether the increased IL-1ra production played a role in modulating the autocrine/paracrine effects of endogenous IL-1β on IL-8 production was not investigated. Unpublished experiments shown in Figure 5.5 would, however, suggest that very high concentrations of IL-1ra added to LPS-stimulated PMN do not influence the production of IL-8 after 18 h, but (as expected), block the effects of IL-1β or IL-1β plus TNFα. Furthermore, other unpublished experiments performed in our laboratory (Fig. 5.6), would also suggest that IL-10 does not enhance the known LPS-stimulated shedding[25a] of the two soluble TNFα receptors (sTNFr),

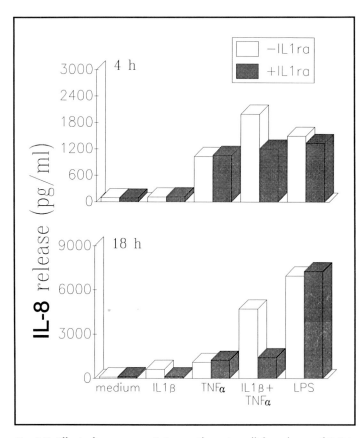

Fig. 5.5. Effect of exogenous IL-1ra on the extracellular release of IL-8 in LPS-, TNFα- and/or IL-1β-stimulated neutrophils.

sTNFr-p55 and sTNFr-p75, respectively. It can be assumed therefore that the LPS-induced release of IL-8 is not negatively balanced by the potential inhibitory action of IL-1ra towards endogenous IL-1β, or by that of sTNFr-p55 or sTNFr-p75 towards endogenous TNFα. In peripheral blood mononuclear cells (PBMC) however, exogenous IL-1ra

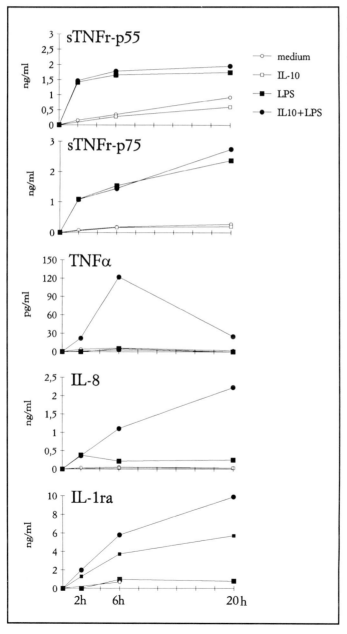

Fig. 5.6. IL-10 does not affect the extracellular release of soluble TNFα receptors in LPS-stimulated neutrophils.

partially inhibited LPS-induced IL-8 synthesis,[26] as well as that induced by IL-2,[27] if used in combination with the sTNFr-p75-Fc fusion protein.

Other groups have later substantially confirmed, and in some instances extended, the aforementioned effects of IL-10 towards cytokine production by neutrophils. Jenkins et al,[28] for example, found that IL-10 (10 ng/ml) significantly reduced the overall IL-1β production, but did not influence that of IL-1ra in response to LPS (10 μg/ml), over a period of 22-24 hr. However, they did not discriminate between the net effects of IL-10 on cell-associated versus secreted IL-1β or IL-1ra.[28] In addition, it was shown that while IL-10 moderately decreased the LPS-induced IL-1β mRNA at times later than 4 h, in agreement with our previous results,[18] it did not affect LPS-induced IL-1ra mRNA levels, in contrast with our findings.[25] We do not have an explanation for this discrepancy at present, but a possible hypothesis is that different concentrations of LPS were used in the two works.[25,28] In any case, the net result of the effects of IL-10 toward neutrophils was again to increase the ratio of IL-1ra to IL-1β, in agreement with the schemes depicted in Figure 3.7.

Very recently, Cavaillon's group[10] reported that IL-10 displayed an extremely high potency to up-regulate the production of IL-1ra induced by TNFα. Conversely, IL-10 was unable to amplify the extracellular production of IL-1ra by LPS (100 ng/ml). They attributed the discrepancy with our results[18] to the fact that we preincubated with IL-10 for 30 min before stimulating neutrophils with LPS. In this regard, it is noteworthy that we observed no significant difference in the effects of IL-10 towards PMN, regardless of whether these cells were preincubated or not with the cytokine.

The capacity of IL-10 to inhibit the LPS-induced extracellular release of TNFα, IL-1β and IL-8 in PMN, was further evidenced by Wang et al,[6] who also reported that IL-10 suppresses the production of IL-1α. In a paper published almost simultaneously, Kasama et al[29] not only confirmed the effect of IL-10 on IL-8 release, but also provided evidence that the release of MIP-1α and MIP-1β by neutrophils stimulated with LPS were dose-dependently inhibited by IL-10. Consistent with our studies,[18] the extracellular production of IL-8, MIP-1α and MIP-1β, as well as the LPS-induced accumulation of mRNA encoding these chemokines, were significantly suppressed by IL-10 only after 4-8 h of culture (Fig. 5.4).[29] It was also shown that in LPS-treated PMN, IL-10 significantly accelerated the decay of IL-8 mRNA,[6,29] as well as that of both MIP-1α/β;[29] these aspects will be further discussed in chapter 6. The results of Kasama et al,[29] together with a later study published by the same group,[30] are in keeping with the possibility that MIP-1α and MIP-1β production in response to LPS are subject to an autocrine regulation similar to that observed for IL-8 (Fig. 3.7).

In one of our most recent papers,[31] we found that IL-10, similarly to its effect on IL-8 production,[18] did not affect the initial phase of Growth Related Gene Product-α (GROα) release stimulated by LPS, but rather exerted its inhibitory effect only after 6 h of incubation (Fig. 5.7). Under these conditions, IL-10 did not affect the LPS-elicited accumulation of GROα mRNA,[31] suggesting that its inhibitory effect on LPS-induced GROα production likely occurred at the translational or post-translational level. Interestingly, IL-10 potentiated the release of GROα induced by TNFα,[31] therefore demonstrating that

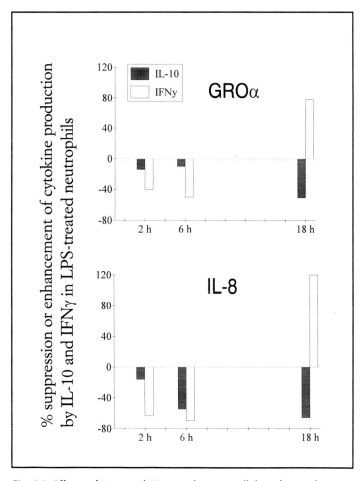

Fig. 5.7. Effects of IL-10 and IFNγ on the extracellular release of GROα and IL-8, in LPS-stimulated neutrophils. The figure depicts the percentage of inhibition or of enhancement by IL-10 and IFNγ of GROα and IL-8 release in LPS-stimulated PMN. Mean values from five independent experiments.

not all the effects of IL-10 towards granulocytes are inhibitory. In fact, as already mentioned, Marie et al[10] also showed that IL-10 potentiates the release of IL-1ra induced by TNFα in PMN. In contrast, the release of GROα triggered by Y-IgG was slightly inhibited by either IL-10 or IFNγ, albeit without a parallel effect at the level of GROα mRNA.[31]

Additional experiments performed in my laboratory were aimed to determine whether the inhibitory effects of IL-10 were selective for cytokine generation. For this purpose, we investigated the effect of 100 U/ml IL-10 on the respiratory burst of PMN incubated for 4 and 18 h in the presence or absence of IFNγ and LPS, which are known to potentiate neutrophil release of reactive oxygen intermediates (ROI).[32,33] As shown in Figure 5.8, and contrary to what has been observed with phorbol-myristate acetate (PMA)-stimulated mouse macrophages,[34] IL-10 failed to significantly influence either the constitutive, or the enhanced ability of PMN to produce superoxide anion (O_2^-) in response to fMLP. Similar results were obtained if TNFα was used as a priming agent instead of IFNγ or LPS, or if phorbol esters were used as a triggering stimulus (not shown). Consistent with these results, IL-10 failed to affect the enhanced expression of heavy chain cytochrome b subunit

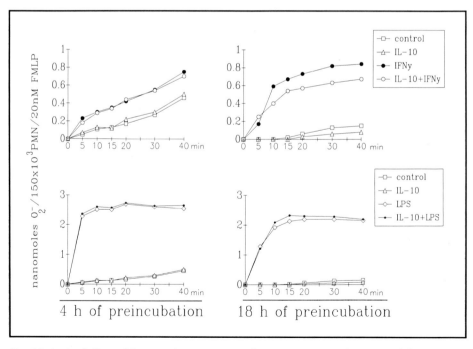

Fig. 5.8. Lack of effect of IL-10 on the potentiation by IFNγ or LPS on the neutrophil respiratory burst. PMN were treated for 4 or 18 h with IL-10, in the presence or absence of IFNγ (100 U/ml) or LPS (1 μg/ml), prior to stimulation with 20 nM fMLP for O_2^- release.

(gp91-phox) mRNA in LPS-treated PMN,[25] which we previously showed to correlate with the IFNγ- or LPS-mediated activation of the neutrophil respiratory burst.[35] Therefore, IL-10 did not influence the release of ROI, independently of the state of neutrophil activation.

All the findings listed above have unveiled important new actions of IL-10 towards leukocytes.[16,36] In particular, they have clearly indicated that IL-10, by regulating PMN cytokine production, may also have an important regulatory role in limiting the duration and extent of the acute inflammatory response. For example, IL-10 has been used recently in rodents to prolong survival in animal models of Gram-negative sepsis provoked by injection of endotoxin.[37-39] Endotoxin-induced toxicity appears to be due to the release of TNFα and/or IL-1, because animals can be also protected from bacterial- and endotoxin-induced shock by neutralization of TNFα and/or IL-1, using either monoclonal antibodies or their physiological antagonists such as IL-1ra.[40,41] The results obtained with neutrophils would suggest that modulation of PMN cytokine production, other than the suppression of monokines, could presumably contribute to the IL-10-mediated protection from lethal endotoxemia. Similarly, an eventual lack of effects of IL-10 on PMN-derived cytokines could help explain the pathogenesis of a human like-inflammatory bowel diseases (IBD) developed by IL-10-deficient mice, generated by targeted mutation.[42] An uncontrolled intestinal immune response, perpetuated by an overproduction of proinflammatory cytokines by macrophages interacting with intestinal-derived bacterial LPS, was invoked as the primary mechanism leading to bowel inflammation in IL-10-deficient mice.[42] It is also possible that the activated neutrophils in the gut of IL-10-deficient mice might be contributing to exaggerated and uncontrolled cytokine production, and thus play an active role in the pathogenesis of gut damage. Finally, IL-10 has been reported to be an important regulatory cytokine during the development of acute lung injury in the rat after intrapulmonary deposition of IgG-immune complexes, characterized by a high number of recruited neutrophils and the detection of high levels of proinflammatory cytokines.[43] Protection by IL-10 of this type of lung injury might act through the inhibition of neutrophil-derived cytokines.

EFFECTS OF INTERFERON-γ (IFNγ)

While IL-4 and IL-10 exert a mainly negative action on the LPS-induced production of proinflammatory cytokines production by PMN, opposite effects have been attributed to IFNγ. The latter is a cytokine produced by T-lymphocytes and NK cells, whose multiple actions in inflammation and on cells of different lineages have been extensively characterized.[44,45] In particular, IFNγ has been shown to modulate functional responses in PMN, such oxidative respiratory burst and degranulation.[46]

At the beginning of 1993, we showed that IFNγ rapidly downregulates the constitutive IL-8 mRNA levels expressed by resting PMN.[24] This action was neither dependent on de novo protein synthesis, nor on an increased rate of IL-8 mRNA degradation.[24] A short preincubation of PMN with IFNγ (30 min) significantly inhibited their release of IL-8 following subsequent stimulation for 2 h with TNFα, fMLP, Y-IgG or LPS. In the same experiments, IFNγ enhanced the respiratory burst of neutrophils induced by TNFα or fMLP,[24] as previously described.[46] Finally, the TNFα-, fMLP- and LPS-induced expression of IL-8 mRNA were also inhibited by IFNγ.[24] In a subsequent study,[47] we observed that if PMN were stimulated for 18 h by LPS, or by TNFα (with or without IL-1β), in the presence of IFNγ, an enhanced yield of IL-8 antigen (as compared to stimuli alone) could be detected in neutrophil cell-free supernatants. We also found that in contrast to IL-8 production the LPS-stimulated extracellular production of TNFα and IL-1β, as well as the TNFα-stimulated production of IL-1β, were markedly enhanced by IFNγ, over the entire period of PMN incubation (up to 18 h).[47] Northern blot analysis showed that all these effects of IFNγ were paralleled by correspondent changes at the level of TNFα, IL-1β and IL-8 mRNA expression.[47] Furthermore, anti-TNFα plus anti-IL-1β monoclonal antibodies blocked the release of IL-8 by PMN incubated with LPS alone or with LPS and IFNγ to a similar extent. The latter findings confirmed that production of endogenous TNFα and IL-1β mediated the augmented release of IL-8 induced by LPS and IFNγ at 18 h (Fig. 3.7).[18,47]

That in neutrophils IFNγ inhibits early mRNA expression and production of IL-8 induced by LPS, while augmenting IL-8 mRNA expression and production at 24 h, was confirmed by Kasama et al,[30] who also described that IFNγ exerted identical actions towards MIP-1α and MIP-1β expression. Moreover, they confirmed that IFNγ enhances several times the production of TNFα and IL-1β triggered by LPS, and that anti-TNFα antibodies, but not anti-IL-1β antibodies, dramatically attenuated the chemokine production from PMN stimulated with LPS and/or IFNγ. Even though Kasama et al[30] did not use neutralizing anti-TNFα and anti-IL-1β antibodies in combination as we did earlier,[18,47] their results also implied (at least in part) that IFNγ augments chemokine expression via an autocrine effect of TNFα.

In another study, Van Dervort and colleagues[48] similarly observed that treatment of human neutrophils with IFNγ and LPS markedly increased the production of TNFα over that of neutrophils treated with LPS alone. The addition of sodium nitroprusside (SNP), an exogenous source of nitric oxide, together with N-acetylcysteine (NAC), which increases the bioavailability of nitric oxide, enhanced even more the IFNγ plus LPS-induced TNFα secretion, over a 20 h incubation. Similarly, exposure of LPS-treated PMN to SNP or SNP plus NAC, but not to NAC alone, increased the production of TNFα compared with exposure to LPS alone, whereas the same exposures were ineffective if carried out in human monocyte-derived macrophages stimulated by LPS, with or without IFNγ.[48] Interestingly, the upmodulation of TNFα production by those two drugs in PMN was not associated with increased amounts of LPS-induced TNFα mRNA,[48] suggesting a posttranscriptional type of regulation. The potential effect of nitric oxide on LPS-induced TNFα production by human neutrophils may represent a paracrine mechanism by which endothelial and vascular smooth muscle cells can augment the activation state of neutrophils as the latter migrate from the blood to sites of infection, and thus enhance the inflammatory response during sepsis. Conversely, these responses could also contribute to the cytokine-mediated tissue injury and organ failure observed during systemic infections.[48]

The actions of IFNγ are not limited to TNFα, IL-1β and IL-8 expression, but involve also that of GROα (Fig. 5.7). We recently found that IFNγ strongly decreases the mRNA levels and production of GROα by PMN incubated with LPS for 2 h, but significantly potentiates GROα-release if PMN are stimulated for 18 h.[31] In contrast, IL-10 did not significantly affect the initial wave of GROα release stimulated by LPS, but started to be inhibitory after 6 h of incubation.[31] Taken together, the effects of IFNγ and IL-10 on LPS-induced GROα were very reminiscent of those exerted by the same cytokines towards the production of IL-8[18,47] (Fig. 5.7). This suggests that the production of GROα induced by LPS is, as already shown for IL-8, MIP-1α and MIP-1β,[18,30] the result of a biphasic response, which would consist of an early phase, directly induced by LPS, followed by a delayed phase, mediated by endogenously produced cytokines such as TNFα, IL-1β, and/or others.

We were also able to show that releases of GROα triggered by TNFα and Y-IgG were, respectively, slightly potentiated or inhibited by IFNγ.[31] Interestingly, modulation by IFNγ of GROα release induced by Y-IgG and TNFα occurred in the absence of changes at the level of GROα mRNA expression. Altogether, the data suggest that LPS, TNFα and Y-IgG induce GROα production through distinct molecular signaling pathways, and that these pathways are differentially sensitive to modulation by IFNγ and IL-10.[31]

In a broader context, the ability of IFNγ to modulated cytokine production by PMN may have important implications in vivo. Using the Shwartzman reaction as a model to investigate the mechanism(s) involved in LPS-induced disease,[49] or using a murine model of endotoxic shock,[50] several investigators have shown that IFNγ appears to be a key mediator in the development and progression of those inflammatory responses. The ability of IFNγ to up-regulate cytokine production, particularly TNFα, IL-1β and IL-8, by neutrophils exposed to provocative doses of LPS, might thus be one of the key factor contributing to the pathogenesis of endotoxic shock.[49] Furthermore, IFNγ has been shown to function as effective prophylactic agent for the infectious complications of chronic granulomatous disease (CGD),[51] but the molecular basis of this effect is not yet well understood.[52] It was in fact initially demonstrated that IFNγ induces a partial correction of NADPH oxidase function in CGD patients, but recent clinical studies have questioned this mechanism since it was seen only in a subset of individuals with a "variant," mild form of CGD. The enhancement of neutrophil-derived cytokine production induced by IFNγ might represent one of the key mechanisms contributing to the improvement of host defense against microbes in CGD patients undergoing IFNγ therapy.

EFFECTS OF INFLAMMATORY MICROCRYSTALS

The deposition of monosodium urate monohydrate (MSU) and calcium pyrophosphate dihydrate (CPPD) microcrystals in joint tissues is associated with the pathogenesis of acute or chronic inflammatory responses such as gout and articular chondrocalcinosis (pseudogout), respectively. One of the characteristics of these diseases is the massive recruitment of neutrophils to these inflammatory sites, whereas mononuclear cells are rarely found in the inflamed joints during the acute phases of these arthropaties.[53]

To better define the mechanisms regulating the pathogenesis of gout and pseudogout, McColl and colleagues investigated the effects of MSU and CPPD alone, or in combination with TNFα and GM-CSF, on cytokine production by neutrophils. As already described in chapter 4, in the course of their studies, McColl's group found that MSU and CPPD induced the generation by human neutrophils of IL-1,[54,55] IL-8 and of very low levels of both GROα[56] and IL-1ra,[54] but not MIP-1α.[56] They observed that treatment of PMN with GM-CSF and TNFα before incubation with suboptimal concentrations of the crystals enhanced the total synthesis of IL-1 induced by microcrystals by 5- to 9-fold, over a 12 h incubation period.[54] Under the same conditions, the levels of total IL-1ra produced by neutrophils were approximately 35 to 43% lower than the expected amounts produced by cytokine-treated cells.[54] As a result of this shift in IL-1 and IL-1ra production by GM-CSF- and TNFα-activated neutrophils, the net biologic activity of IL-1 secreted in response to the microcrystals was enhanced. Interestingly, colchicine (10 μM) did not significantly affect the production of IL-1 and IL-1ra by either unstimulated PMN or by PMN incubated in the presence of GM-CSF, TNFα, or crystals. However, treatment of neutrophils with colchicine prior to incubation with GM-CSF or TNFα inhibited the crystal-induced IL-1 production by 50 to 55%, but failed to affect that of IL-1ra. In this situation therefore, the IL-1ra to IL-1 ratio increased significantly by 185-220%.[54] Altogether, these results not only demonstrated that IL-1 and IL-1ra production by human neutrophils were differentially regulated, but that the combined presence of GM-CSF or TNFα and microcrystals favored the production of biologically active IL-1 over that of IL-1ra, thereby potentially amplifying the inflammatory response associated with crystal-induced diseases. In addition, because colchicine selectively inhibited IL-1 without affecting IL-1ra production, they uncovered another putative site of action of the drug.

More recently, the same group also showed that IL-8 production induced by TNFα and GM-CSF was synergistically enhanced in the presence of MSU or CPPD, whereas MIP-1α secretion induced by TNFα was completely inhibited in the presence of either MSU or CPPD.[56] The latter results are of a great relevance because they suggest that the combination of MSU and CPPD with TNFα and GM-CSF leads to the production of IL-8 by neutrophils and abolishes the release of MIP-1α. If these findings reflect the in vivo situation, such a regulation by MSU and CPPD could have important implications, because they suggest that neutrophils might not only contribute to their own recruitment, but also that they might prevent the recruitment of mononuclear cells to the synovial environment,[56] in accordance with the pathological state of gout and pseudogout, where the predominant inflammatory cell is the neutrophil.

EFFECTS OF OTHER SUBSTANCES

The glucocorticoid, Dexamethasone (DEX), has been shown to exert numerous modulatory effects towards the production of cytokine by neutrophils. We have examined the action of Dexamethasone (DEX) on the expression of IL-8 mRNA in control and IFNγ-treated neutrophils.[57] We found that DEX alone inhibited the constitutive mRNA expression of IL-8, and potentiated the suppression mediated by IFNγ.[57] In contrast to that of IFNγ, the negative action of DEX required at least 90 min to occur, and was abrogated by cycloheximide.[57] This is not surprising in the light of the known effect of DEX on new protein synthesis in PMN.[58] Our results also indicated that DEX and IFNγ down-regulate IL-8 mRNA expression by different pathways.[57] Wertheim et al[5] extended our observations and reported that similar to IL-4, DEX downregulated IL-8 mRNA and protein secretion by LPS-stimulated PMN.

In keeping with its anti-inflammatory properties,[59] DEX was also shown to induce the release of the IL-1RII.[60] IL-13 was shown to do the same,[61] in addition to inducing the intracellular and soluble IL-1ra isoforms from PMN.[62] In this context, some very recent findings on the ability of DEX to prolong neutrophil survival in vitro[63,64] are worth mentioning, as they open new avenues in interpreting the role of corticosteroids in inflammation.

Finally, treatment of PMN with (prostaglandin E_2) PGE_2 failed to inhibit the LPS-induced mRNA expression and generation of IL-8.[5] Intriguingly, these findings are similar to what has been observed in alveolar macrophages[65], but contrast with the PGE_2-elicited down-regulation of IL-8 production in monocytes.[66]

REFERENCES

1. In: Coffey RG, ed. Granulocyte Responses to Cytokines: Basic and Clinical Research. New York: Marcel Dekker, 1992:1-721.
2. Paul WP. Interleukin-4: a prototypic immunoregulatory lymphokine. Blood 1991; 77:1859-1870.
3. Finkelman FD, Urban JF, Paul WP et al. In: Spits H, ed. IL-4: Structure and Function. Boca Raton: CRC Press, 1992:33-54.
4. Boey H, Rosenbaum R, Castracane J et al. Interleukin-4 is a neutrophil activator. J Allergy Clin Immunol 1989; 83:978-984.
5. Wertheim WA, Kunkel SL, Standiford TJ et al. Regulation of neutrophil-derived IL-8: the role of prostaglandin E_2, dexamethasone, and IL-4. J Immunol 1993; 151:2166-2175.
6. Wang P, Wu P, Anthes JC et al. Interleukin-10 inhibits Interleukin-8 production in human neutrophils. Blood 1994; 83:2678-2683.
7. Cavaillon JM, Marie C, Pitton C et al. Regulation of neutrophil derived IL-8 production by anti-inflammatory cytokines (IL-4, IL-10 and TGFβ). In: Faist E, ed. Proceedings 3rd International Congress on the Immune Consequences of Trauma Shock and Sepsis. Munich: Pabst Science Publishers, 1996, in press.
8. Malyak M, Smith MF, Abel AA et al. Peripheral blood neutrophil production of IL-1ra and IL-1β. J Clin Immunol 1994; 14:20-30.
9. Re F, Mengozzi M, Muzio M et al. Expression of Interleukin 1 receptor antagonist by human circulating polymorphonuclear cells. Eur J Immunol 1993; 23:570-573.
10. Marie C, Pitton C, Fitting C et al. IL-10 and IL-4 synergize with TNFα to induce IL-1ra production by human neutrophils. Cytokine 1996, in press.
11. Colotta F, Re F, Muzio M et al. Interleukin-1 type II receptor: a decoy target for IL-1 that is regulated by IL-4. Science 1993; 261:472-475.
12. Colotta F, Re F, Polentarutti N et al. Modulation of granulocyte survival and programmed cell death by cytokines and bacterial products. Blood 1992; 80:2012-2020.
13. Sims JE, Gayle MA, Slack JL et al. Interleukin-1 signaling may occur exclusively via the type I receptor. Proc Natl Acad Sci USA 1993; 90:6155-6159.
14. Colotta F, Dower SK, Sims JE et al. The type II "decoy" receptor: novel regulatory pathway for interleukin-1. Immunol Today 1994; 15:562-566.
15. Colotta F, Orlando S, Fadlon EJ et al. Chemoattractants induce rapid release of the interleukin 1 type II decoy receptor in human polymorphonuclear cells. J Exp Med 1995; 181:2181-2188.
16. Moore KW, O'Garra A, de Waal Malefyt R et al. Interleukin 10. Ann Rev Immunol 1993; 11:165-190.
17. Takanaski S, Nonaka R, Xing Z et al. Interleukin-10 inhibits lipopolysaccharide-induced survival and cytokine production by human peripheral blood eosinophils. J Exp Med 1994; 180:711-715.

18. Cassatella MA, Meda L, Bonora S et al. Interleukin-10 inhibits the release of proinflammatory cytokines from human polymorphonuclear leukocytes. Evidence for an autocrine role of TNFα and IL-1β in mediating the production of IL-8 triggered by lipopolysaccharide. J Exp Med 1993; 178:2207-2211.
19. Donnelly RP, Freeman SL, Hayes MP. Inhibition of IL-10 expression by IFNγ up-regulates transcription of TNFα in human monocytes. J Immunol 1995; 155:1420-1427.
20. Bogdan C, Paik J, Vodovotz Y et al. Contrasting mechanisms for suppression of macrophage cytokine release by transforming growth factor-β and interleukin-10. J Biol Chem 1992; 267:23301-23308.
21. Strieter RM, Kasahara K, Allen RM et al. Cytokine-induced neutrophil-derived Interleukin-8. Am J Pathol 1992; 141:397-407.
22. Fujishima S, Hoffman AR, Vu T et al. Regulation of neutrophil interleukin 8 gene expression and protein secretion by LPS, TNFα and IL-1β. J Cell Physiol 1993; 154:478-485.
23. DeForge LE, Kenney JS, Jones ML et al. Biphasic production of IL-8 in lipopolysaccharide (LPS)-stimulated human whole blood. Separation of LPS- and cytokine-stimulated components using anti-tumor necrosis factor and anti-IL-1 antibodies. J Immunol 1992; 148:2133-2141.
24. Cassatella MA, Guasparri I, Ceska M et al. Interferon-γ inhibits Interleukin-8 production by human polymorphonuclear leukocytes. Immunology 1993; 78:177-184.
24a. Cassatella MA, Bazzoni F, Ceska M et al. Interleukin-8 production by human polymorphonuclear leukocytes. The chemoattractant Formyl-Methionyl-Leucyl-Phenylalanine induces the gene expression and release of Interleukin 8 through a pertussis toxin sensitive pathway. J Immunol 1992; 148:3216-3220.
25. Cassatella MA, Meda L, Gasperini S et al. Interleukin 10 up-regulates IL-1 receptor antagonist production from lipolysaccharide-stimulated human polymorphonuclear leukocytes by delaying mRNA degradation. J Exp Med 1994; 179:1695-1699.
25a. van der Poll T, Calvano SE, Kumar A et al. Endotoxin induces downregulation of tumor necrosis factor receptors on circulating monocytes and granulocytes in humans. Blood 1995; 86:2754-2759.
26. Porat B, Poutsiaka DD, Miller LC et al. Interleukin-1 (IL-1) receptor blockade reduces endotoxin and *Borrelia burgdorferi*-stimulated IL-8 synthesis in human mononuclear cells. FASEB J 1992; 6:2482-2486.
27. Tilg H, Shapiro L, Atkins MB et al. Induction of circulating and erythrocyte-bound IL-8 by IL-2 immunotherapy and suppression of its in vitro production by IL-1 receptor antagonist and soluble tumor necrosis factor receptor (p75) chimera. J Immunol 1993; 151:3299-3307.
28. Jenkins JK, Malyak M, Arend, WP. The effects of Interleukin-10 on Interleukin-1 receptor antagonist and Interleukin-1β production in human monocytes and neutrophils. Lymphokine Cytokine Res 1994; 13:47-54.

29. Kasama T, Strieter RM, Lukacs NW et al. Regulation of neutrophil-derived chemokine expression by IL-10. J Immunol 1994; 152:3559-3569.
30. Kasama T, Strieter RM, Lukacs NW et al. Interferon gamma modulates the expression of neutrophil-derived chemokines. J Invest Med 1995; 43:58-67.
31. Gasperini S, Calzetti F, Russo MP et al. Regulation of GROα production in human granulocytes. J Inflamm 1995; 45:143-151.
32. Berton G, Cassatella MA. Modulation of neutrophil functions by interferon gamma. In: Coffey RG, ed. Granulocytes Responses to Cytokine: Basic and Clinical Researches. New York: Marcel Dekker Inc., 1992: 437-456.
33. Guthrie LA, McPhail LC, Henson PM et al. Priming of neutrophils for enhanced release of oxygen metabolites by bacterial lipopolysaccharide. J Exp Med 1984; 160:1656-1671.
34. Bogdan C, Vodovotz Y, Nathan C. Macrophage deactivation by Interleukin 10. J Exp Med 1991; 174:1549-1555.
35. Cassatella MA, Bazzoni F, Flynn RM et al. Molecular basis of Interferon-γ and Lypopolysaccharide enhancement of phagocyte respiratory burst capability. Studies on the gene expression of several NADPH oxidase components. J Biol Chem 1990; 265:20241-20246.
36. Mosmann, TR. Properties and functions of Interleukin-10. Adv Immunol 1994; 56:1-26.
37. Gerard C, Bruyns C, Marchant A et al. Interleukin 10 reduces the release of tumor necrosis factor and prevents lethality in experimental endotoxemia. J Exp Med 1993; 177:547-550.
38. Howard M, Muchamuel T, Andrade S et al. Interleukin 10 protects mice from lethal endotoxemia. J Exp Med 1993; 177:1205-1208.
39. Standiford TJ, Strieter RM, Lukacs NW et al. Neutralization of IL-10 increases lethality in endotoxemia. Cooperative effects of macrophage inflammatory protein-2 and tumor necrosis factor. J Immunol 1995; 155:2222-2229.
40. Vassalli P. The pathophysiology of tumor necrosis factors. Ann Rev Immunol 1992; 10:411-452.
41. Ohlsson K, Bjork P, Bergenfeldt M et al. An interleukin 1 receptor antagonist reduces mortality in endotoxin shock. Nature 1990; 348:550-553.
42. Kuhn R, Lohler J, Rennick D et al. Interleukin-10-deficient mice develop chronic enterocolitis. Cell 1993; 75:263-274.
43. Shanley TP, Schmal H, Friedl HP et al. Regulatory effects of IL-10 of intrinsic IL-10 in IgG immune complex-induced lung injury. J Immunol 1995; 154:3454-3460.
44. Vilcek J, Gray PW, Rinderknecht E et al. Interferon-γ, a lymphokine for all seasons. Lymphokines 1985; 11:1-32.
45. Farrar MA, Schreiber RD. The molecular cell biology of Interferon-γ and its receptor. Ann Rev Immunol 1993; 11:571-611.
46. Berton G, Zeni L, Cassatella MA et al. Gamma interferon is able to enhance the oxidative metabolism of human neutrophils. Biochem Byophis

Res Commun 1986; 138:1276-1282.
47. Meda L, Gasperini S, Ceska M et al. Modulation of proinflammatory cytokine release from human polymorphonuclear leukocytes by gamma interferon. Cell Immunol 1994; 57:448-461.
48. Van Dervort AL, Yan L, Madara PJ et al. Nitric oxide regulates endotoxin-induced TNFα production by human neutrophils. J Immunol 1994; 152:4102-4109.
49. Heremans HR, Van Damme J, Dillen C et al. Interferon gamma, a mediator of lethal lipopolysaccharide-induced Shwartzman-like shock reactions in mice. J Exp Med 1990; 171:1853-1869.
50. Heinzel FP. The role of IFNγ in the pathology of experimental endotoxiemia. J Immunol 1990; 145:2920-2924.
51. Ezekowitz RAB and the international chronic granulomatous disease cooperative study group. A controlled trial of interferon gamma to prevent infection in chronic granulomatous disease. N Engl J Med 1991; 324:509-516.
52. Malech HL. Phagocyte oxidative mechanisms. Curr Op Hematol 1993; 1:123-132.
53. Terkeltaub R. Gout: crystal-induced inflammation. In: Gallin JI, Goldstein IM, Snyderman R. Inflammation: Basic Principles and Clinical Correlates. 2nd edition. New York: Raven Press, 1992: 977-981.
54. Roberge CJ, de Medicis R, Dayer JM et al. Crystal-induced neutrophil activation. V. Differential production of biologically active IL-1 and IL-1 receptor antagonist. J Immunol 1994; 152:5485-5494.
55. Roberge CJ, Grassi J, de Medicis R et al. Crystal-neutrophil interactions lead to IL-1 synthesis. Agents Actions 1991; 34:38-41.
56. Hachicha M, Naccache PH, McColl SR. Inflammatory mycrocrystal differentially regulate the secretion of macrophage Inflammatory protein 1 and interleukin 8 by human neutrophils: a possible mechanism of neutrophil recruitment to sites of inflammation in synovitis. J Exp Med 1995; 182:2019-2025.
57. Cassatella MA, Aste M, Calzetti F et al. Studies on the regulatory mechanisms of Interleukin-8 gene expression in resting and IFNγ-treated neutrophils. Evidence on the capability of staurosporine of inducing the production of IL-8 by human neutrophils. Biochem Biophys Res Commun 1993; 190:660-667 [published erratum in 1993; 192:324].
58. Blowers LE, Jayson MIV, Jasani MK. Effect of dexamethasone on polypeptides synthesized in polymorphonuclear leukocytes. FEBS Lett 1985; 181:362-366.
59. Cupps TR, Fauci AS. Corticosteroid-mediated immunoregulation in man. Immunol Rev 1982; 65:133-155.
60. Re F, Muzio M, De Rossi M et al. The type II receptor as a decoy target for IL-1 in polymorphonuclear leukocytes: characterization of induction by dexamethasone and ligand binding properties of the released receptor. J Exp Med 1994; 179:739-743.

61. Colotta F, Re F, Muzio M et al. Interleukin-13 induces expression and release of the IL-1 decoy receptor in human polymorphonuclear cells. J Biol Chem 1994: 269:12403-12406.
62. Muzio M, Re F, Sironi M et al. Interleukin-13 induces the production of Interleukin-1 receptor antagonist (IL-1ra) and the expression of the mRNA for the intracellular (keratinocyte) form of IL-1ra in human myelo-monocytic cells. Blood 1994; 83:1738-1743.
63. Cox G. Glucocorticoid treatment inhibits apoptosis in human neutrophils. Separation of survival and activation outcomes. J Immunol 1995; 154:4719-4725.
64. Liles WC, Dale DC, and Klebanoff SJ. Glucocorticoids inhibit apoptosis of human neutrophils. Blood 1995; 86:3181-3188.
65. Standiford TJ, Kunkel SL, Rolfe MW et al. Regulation of human alveolar macrophage and blood monocyte-derived IL-8 by PGE_2 and dexamethasone. Am J Respir Cell Mol Biol 1992; 6:75-81.
66. Strieter RM, Remick DG, Lynch III JP et al. Differential regulation of TNFα in human alveolar macrophage and peripheral blood monocytes: a cellular and molecular analysis. Am J Respir Cell Mol Biol 1989; 1:57-63.

CHAPTER 6

MOLECULAR REGULATION OF CYTOKINE PRODUCTION IN NEUTROPHILS

The phenotypic differences that distinguish the various kinds of cells in higher eukaryotes are merely due to an overall different pattern of gene expression. Gene expression is not an automatic process, but it can be regulated in a gene-specific manner. There are many steps in the pathway leading from DNA to protein, and, in principle, each of them can be regulated. Thus, a cell can control the proteins it makes by: (1) controlling when and how often a given gene is transcribed (transcriptional control); (2) controlling how the primary transcript is spliced or otherwise processed (RNA processing control); (3) selecting which completed mRNAs in the cell nucleus are exported to the cytoplasm (RNA transport control); (4) selectively destabilizing certain molecules in the cytoplasm (mRNA stability control); (5) selecting which mRNAs in the cytoplasm are translated by ribosomes and with what efficiency (translational control); and (6) selectively compartmentalizing, activating, or inactivating specific protein molecules after they have been made (protein stability control). All the regulatory mechanisms that control gene expression following transcription are referred to as post-transcriptional control. For most genes, the initiation of RNA transcription is the most important point of control. However, post-transcriptional forms of control are also crucial for many genes to modulate the amount of product that is made. In general, every step in gene expression that could be controlled is in principle likely to be regulated under some circumstances.

It is well established that cytokine gene expression is also regulated at several levels, including modulation of RNA processing, changes in the stability of mature mRNA or in the efficiency of mRNA

Cytokines Produced by Polymorphonuclear Neutrophils: Molecular and Biological Aspects, by Marco A. Cassatella. © 1996 R.G. Landes Company.

translation, but a large body of evidence indicates that transcriptional control is the predominant type of regulation. Cytokine genes are usually transcriptionally inactive and need a specific stimulus to overcome this repression.

An increasing body of knowledge is now available on the molecular mechanisms which regulate cytokine gene expression in myelomonocytic cells, including neutrophils. As in monocytes or macrophages (which have been more extensively studied in this regard), the control of cytokine gene expression in neutrophils can take place at the transcriptional, post-transcriptional, translational and post-translational levels. This evidence is mostly derived most from direct experimental demonstrations, but also from indirect indications obtained through the use of selective metabolic inhibitors. What follows is a brief overview of what is known on this topic. The reader will notice that our knowledge on polymorphonuclear leukocytes (PMN) is for the moment rather limited, especially due to the difficulty or the impossibility, sometimes, to extend to neutrophils many of the new, sophisticated recombinant DNA techniques that are utilized in other cells.

EFFECTS OF METABOLIC INHIBITORS

A straightforward approach taken by many investigators to obtain information on the mechanisms involved in gene regulation, is to examine the effects of metabolic inhibitors. A class of widely used compounds are the inibitors of protein synthesis, such as cycloheximide (CHX), puromycin and emetine, with CHX being the most widely used. Table 6.1 reports the effects of CHX on neutrophil expression of cytokine mRNA reported to date.

In resting PMN, CHX by itself has been shown to either induce the mRNA for IL-1β,[1] IL-1ra,[2-4] TNFα[5] and MIP-1α,[6,7] or to increase the basal mRNA expression of IL-1ra[2-4] and IL-8.[7-11] Cicco et al[12] also reported an induction of IL-6 mRNA by CHX, but we could not reproduce that effect (our unpublished results). Altogether, these results indicate a superinduction phenomenon by CHX, presumably because the drug reduces the presence of labile inhibitor(s) of IL-1β, IL-1ra, MIP-1α, TNFα, and IL-8 transcription, or because labile nuclease(s) degrade the related cytokine mRNAs.

In IL-2-stimulated PMN, the induction of TNFα and IL-8 mRNAs was not found to be influenced by CHX.[5,13] Conversely, CHX has been observed to superinduce IL-1β mRNA by IL-1β;[1] IL-1ra mRNA by LPS,[2] GM-CSF,[2] IL-4,[2] IL-13,[3] and TGFβ$_1$;[4] TNFα mRNA by IL-8 or heat-killed *C.albicans*;[5] MIP-1α/β mRNAs by LPS in the presence or absence of GM-CSF or IL-10;[6,7] and of IL-8 mRNA by LPS,[7,11] IL-8 or heat-killed *Candida*.[5] All the latter findings indicate that the induction of all these cytokine mRNAs does not require de novo protein synthesis. The superinduction of cytokine mRNA by CHX is thought to be due, as already mentioned, to unstable transcriptional repressors

Table 6.1. Effect of cycloheximide on cytokine mRNA expression in human neutrophils

Cytokine mRNA	Experimental Conditions	Effect of Cycloheximide
IL-1β	Resting	↑[1]
	TNFα	↓[1]
	IL-1β	↑[1]
	TNFα + IL-1β	↓[1]
IL-1ra	Resting	↑[2-4]
	LPS	↑[2]
	GM-CSF	↑[2]
	IL-4	↑[2]
	IL-13	↑[3]
	TGFβ$_1$	↑[4]
IL-8	Resting on plastic	↑[8]
	Resting	↑[7-11]
	LPS	↑[7,11]
	IL-10	↑[7]
	LPS + IL-10	↑[7]
	IL-2	=[13]
	GM-CSF	↓[13]
	IL-8	↑[13]
	C. albicans	↑[13]
	IFNγ	=[9]
TNFα	Resting	=[5]
	IL-2	=[5]
	GM-CSF	↓[5]
	IL-8	↑[5]
	C. albicans	↑[5]
MIP-1α	Resting	↑[6,7]
	LPS	↑[6,7]
	GM-CSF	↑[6]
	LPS + GM-CSF	↑[6]
	IL-10	↑[7]
	LPS + IL-10	↑[7]
MIP-1β	Resting	↑[7]
	LPS	↑[7]
	IL-10	=[7]
	LPS + IL-10	↑[7]

= : no detectable effect.
↑ : upregulatory effect.
↓ : downregulatory effect.

or to the decay of unstable ribonucleases. Surprisingly, the negative effect of IFNγ on the early accumulation of IL-8 mRNA in PMN was not prevented by CHX pretreatment, even though CHX by itself caused an accumulation in the increase of IL-8 mRNA.[9] It was as if IFNγ prevented the superinductory effect of CHX on IL-8 mRNA expression. In contrast, the induction of IL-1β mRNA by TNFα,[1] and of TNFα and IL-8 by GM-CSF,[5,13] were found to be abrogated by CHX, suggesting a requirement for the synthesis of regulatory protein(s) that are not constitutively present in the PMN. A careful analysis of the data reported in Table 6.1 makes it clear that the influence of de novo protein synthesis towards the steady-state level of a given cytokine mRNA can vary considerably depending upon the triggering stimulus.

Certain proto-oncogenes and cytokines,[14,15] including for example most Interleukins, TNFα[15] and hematopoietic growth factors,[14] which are usually expressed transiently, contain one or several AUUUA-rich sequences in the 3'-untranslated region(s). These unique sequences may represent recognition sites for mRNA processing pathways which degrade specific transcripts.[16] For instance, removal of the AUUUA repeats can lead to stabilization of the mRNA, and conversely, insertion of these segments into a stable mRNA can lead to its destabilization. Strikingly, these AUUUA-rich sequences in the 3'-untranslated regions have not been identified in genes known to express stable levels of transcripts.[14,15] The mechanisms by which the AUUUA motifs destabilize mRNA, however, remain poorly understood.[14] A variety of investigators have shown that translation and mRNA turnover are often coupled. For example, Shaw and Kamen have demonstrated that GM-CSF mRNA is very unstable, with a half-life of less than 30 min.[17] They found that when GM-CSF gene expression was induced, the corresponding mRNA was stabilized, in that its half-life was greater than 2 h. Furthermore, when they treated cells that contained these unstable mRNAs with CHX, the mRNAs were stabilized. Thus, they concluded that translation of these AUUUA-rich mRNAs was coupled to degradation. The sensitivity of these pathways to CHX has further suggested that ribonucleases may be responsible for the instability of certain transcripts.

Another metabolic inhibitor that is widely used is Actinomycin D (ACT D), a blocker of RNA synthesis, which can yield indirect evidence that the induction of a given gene is regulated at the transcriptional level. The effects of ACT D towards the expression of various cytokine mRNAs in resting and stimulated neutrophils are summarized in Table 6.2. Preincubation of PMN with ACT D leads to a marked reduction of the constitutive IL-1ra[2-4] and IL-8 mRNA expression.[7,8,11] In addition, ACT D strongly blocks the induction of IL-1ra mRNA by LPS,[2] GM-CSF,[2] IL-13,[3] and TGFβ$_1$,[4] that of IL-8 mRNA by LPS, regardless of the presence of IL-10,[7,11] and that of MIP-1α/β mRNAs by LPS, independently of the presence of GM-CSF or IL-10.[6,7]

Table 6.2. Inhibitory effects of Actinomycin D on cytokine mRNA expression in human neutrophils

Cytokine mRNA	Experimental Conditions	References
IL-1β	Resting	11
IL-1ra	Resting	2, 3, 4
	LPS	2
	GM-CSF	2
	IL-4	2
	IL-13	3
	TGFβ$_1$	4
IL-8	Resting on plastic	8
	Resting	7, 11
	LPS	7, 11
	IL-10	7
	LPS + IL-10	7
TNFα	Resting	5
MIP-1α	LPS	6, 7
	LPS + GM-CSF	6
	LPS + IL-10	7
MIP-1β	LPS	7
	LPS + IL-10	7

Figure 6.1 shows a Northern blot experiment in which neutrophils were stimulated for 1 h with LPS (1 µg/ml) or formyl-methionyl-leucyl-phenylalanine, (fMLP)(10 nM) after a 20 min preincubation with ACT D. Apart from its inhibitory effect towards LPS-induced cytokine mRNA accumulation, ACT D also inhibits the fMLP-dependent induction of IL-8 and IL-1β mRNA. Altogether, results obtained with ACT D have indirectly suggested that the induction of IL-1β, IL-1ra, IL-8 and MIP-1α transcripts under these conditions could involve transcriptional events, and (as it will be outlined below and summarized in Table 6.3) it is already known to be the case for IL-1β, IL-8 and MIP-1α.[1,11,18]

TRANSCRIPTIONAL AND POST-TRANSCRIPTIONAL REGULATION

Nuclear run-on (or run-off) assays allow a direct demonstration of the regulation of a given gene at the level of transcription. However, very few run-on experiments performed in PMN have been reported to date, and have involved only IL-1β, IL-8 and MIP-1α genes.[1,11,18] This probably owes to the extreme difficulties encountered in attempting to detect the very low transcriptional activity of neutrophils.[19,20] In our experience, for instance, the transcriptional rates of p47-phox,[21] p22-phox,[21] FcγR-I,[21] IL-8,[11] IL-1β[11] and MIP-1α[18] genes were found

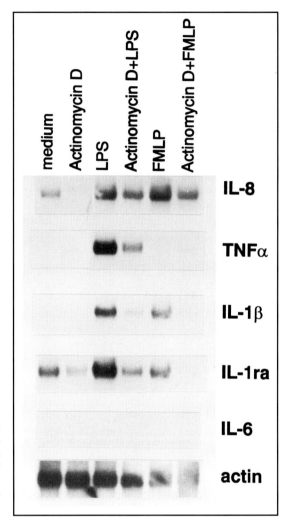

Fig. 6.1. Effect of Actinomycin D on cytokine mRNA accumulation in neutrophils stimulated with LPS or fMLP. PMN were stimulated for 60 min, and then subjected to Northern blot analysis.

to be at least two-to-three orders of magnitude lower in neutrophils than those observed in an equivalent number of myeloid leukemic cell lines or of Natural Killer (NK) cells, for other genes.[22,23] Suffice it to say that in order to detect a radioactive signal from neutrophil nuclear RNA, after its hybridization to filter bound cDNAs we needed to expose the autoradiographic films (using intensifying screens) for at least 40 days at -70°, whereas, with other cell types, a very good signal is usually detected after an overnight exposure only!

The first group who described the transcriptional regulation of a cytokine gene in neutrophils was that of Marucha et al.[1] After having demonstrated that IL-1 and TNFα rapidly (but transiently) induced

Table 6.3. Molecular regulation of cytokine production in human neutrophils

CYTOKINE	EXPERIMENTAL CONDITIONS	EFFECT ON EXPRESSION	MECHANISM (REF) TRANSCRIPTION	mRNA STABILITY	TRANSLATION	POST-TRANSLATION
IL-1β	IL-1 +/- TNFα	↑	1	1		
	LPS	↑	11		34	
IL-8	IL-4 + LPS	↓*			36	36
	LPS	↑	11			
	IFNγ + LPS	↓*	11			
	IL-10 + LPS	↓*	11	7, 20		
	Y-IgG	↑**				26
	IFNγ + LPS/TNFα	↑**				26
	CHX	↑				41
MIP-1α	LPS	↑	18			
	IFNγ + LPS	↓*	18	28		
	IL-10 + LPS	↓*		7		
	GM-CSF + LPS	↑*		6		
	CPPD/MSU + TNFα	↓***			29	
MIP-1β	IL-10 + LPS	↓*		7		
	IFNγ + LPS	↑*		28		
IL-1ra	IL-13	↑		3		
	TGFβ1			4		
	IL-10 + LPS	↑*		25		
	CPPD + TNFα	↓***			38	
GROα	Y-IgG	↑				39
TNFα	GM-CSF	↑				43

Y-IgG: Saccaromyces cerevisiae opsonized with IgG; CHX: cycloheximide; MSU: monosodium urate; CPPD: calcium pyrophosphate dihydrate; *: relative to LPS; **: relative to LPS or TNFα; ***: relative to TNFα.

IL-1β mRNA,[24] they further investigated whether this induction was regulated at the molecular level.[1] Overcoming the various technical difficulties involved in performing nuclear run-on analyses in PMN, they were able to demonstrate that IL-1, TNFα, and IL-1 plus TNFα induced within 1 h the transcription of the IL-1β gene by 33-, 61-, and 99-fold, respectively, and that by 2 h, the levels of IL-1β transcription were dramatically reduced.[1] These relative rates of transcription correlated well with the relative levels of IL-1β steady-state mRNA induced by the two stimuli.[24] By the same technique, they were also able to show that CHX had no detectable effect on the induction of transcription of the IL-1β gene in resting cells, and that it also did not block the induction of transcription of IL-1β by TNFα.[1] These findings supported the conclusion that induction of IL-1β gene by TNFα occurs via a signal transduction pathway that does not require de novo protein synthesis, and that is likely mediated through the activation of pre-existing transcription factors. In addition, these investigators found that post-transcriptional mechanisms could also regulate IL-1β gene expression in PMN. For this purpose, they determined the IL-1β mRNA stability in IL-1-, TNFα- and IL-1 plus TNFα-treated PMN. When steady-state IL-1β mRNA levels were at their peak (45 to 90 min after stimulation with cytokines), the calculated IL-1β mRNA half-life was approximately 1.5 h for TNFα and TNFα plus IL-1β-treated PMN, and approximately 1 h for IL-1β-treated PMN. Interestingly, after 2 h of cytokine treatment, IL-1β mRNA degradation had increased to yield a half-life of about 18 min. Because the initial mRNA half-life from cytokine-treated PMN was higher than the secondary half-life, the authors concluded that IL-1, TNFα, and TNFα plus IL-1 can modulate the stability of IL-1β mRNA,[1] in addition to inducing the transcriptional activation of the IL-1β gene.

That the IL-1β gene can be regulated at the transcriptional level in PMN has been also confirmed in my laboratory[11] (Fig. 6.2). We indeed showed that in neutrophils cultured for 4 h, the constitutive transcriptional activity of the IL-1β gene is almost undetectable, but it can be strongly induced by LPS[11] (Fig. 6.2). Interestingly, at that time point, the up-regulatory effect of LPS on IL-1β gene transcription was neither significantly modified by IFNγ,[11] nor by IL-10 (Fig. 6.2). Moreover, under identical experimental conditions, both IFNγ (our unpublished observations) and IL-10[25] (Fig. 6.3) failed to significantly affect the stability of IL-1β mRNA isolated from LPS-treated PMN. Although IFNγ[26] and IL-10[25,27] are known to increase and decrease, respectively, the steady-state IL-1β mRNA levels in LPS-treated PMN, it is not possible to conclude that these two cytokines are ineffective towards IL-1β gene transcription or IL-1β mRNA stability, until more extended transcriptional and post-transcriptional analyses of IL-1β gene expression will be performed.

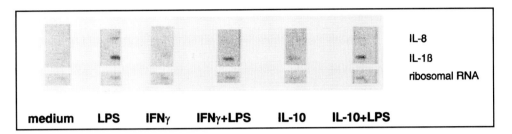

Fig. 6.2. Nuclear run-on assays of IL-8 and IL-1β gene transcription in human neutrophils cultured for 4 h in the presence or absence of IFNγ, IL-10, and/or LPS.

In the same study, we also showed that human granulocytes can actively transcribe the IL-8 gene.[11] We were also very interested in investigating the molecular basis of the early inhibitory action of IFNγ on LPS-induced IL-8 gene expression.[9,28] In this regard, our nuclear run-on analysis revealed that LPS induced the transcription of the IL-8 gene in PMN stimulated for 4 h, and that IFNγ markedly inhibited the rate of this LPS-activated IL-8 gene transcription[11] (Fig. 6.2). Transcriptional inhibition also appeared to be the only mechanism by which IFNγ inhibits the early gene expression in PMN. We,[11] as well as others,[28] found that IFNγ does not affect the IL-8 mRNA stability in PMN treated with LPS for up to 24. Further supporting the lack of effect of IFNγ towards IL-8 mRNA stability in PMN, was an earlier finding by my group regarding the down-modulatory action of IFNγ on the constitutive IL-8 mRNA levels expressed in resting PMN, which was not accounted for by an increased rate of degradation of IL-8 mRNA.[9]

The mechanisms underlying the inhibitory actions of IL-10 towards IL-8 mRNA accumulation in LPS-treated neutrophil have also been extensively analyzed. Our experiments indicate that IL-10 inhibits the rate of LPS-stimulated IL-8 gene transcription in PMN (Fig. 6.2). Kasama et al,[7] as well as Wang et al,[20] reported, instead, that the inhibitory effect of IL-10 towards LPS-induced IL-8 mRNA accumulation correlated with an enhancement of IL-8 mRNA degradation. However, the latter two groups did not show data on the transcriptional rate of IL-8 gene. Nevertheless, it can be envisaged that IL-10 inhibits the LPS-induced IL-8 mRNA accumulation through both inhibition of IL-8 gene transcription and enhanced degradation of IL-8 mRNA.

Another cytokine gene which can be regulated at the level of both transcription and mRNA stability in PMN is the MIP-1α gene.[18,28] As discussed in chapter 4, LPS is a potent inducer of MIP-1α mRNA accumulation and production in human PMN.[6,7,18,28] Neutrophil

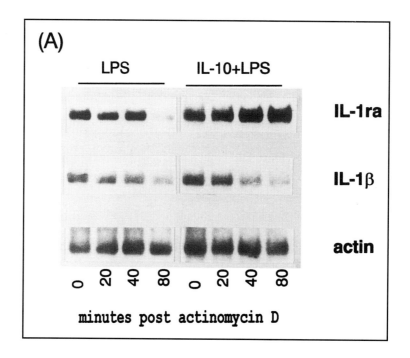
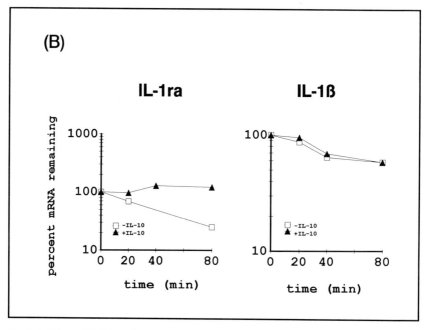

Fig. 6.3. Effect of IL-10 on the turnover rate of IL-1ra and IL-1β mRNAs in LPS-stimulated neutrophils. (A, top) Neutrophils were cultured with LPS, in the presence or absence of IL-10 for 3 h and 30 min, and then treated with actinomycin D prior to Northern blot analysis. (B, bottom) RNA bands were analyzed densitometrically, and the resulting values plotted against time.

pretreatment with IL-10 significantly inhibits the expression of MIP-1α,[7] while IFNγ exerts a biphasic effect, as also observed for IL-8[9,28] and MIP-1β.[28] IFNγ inhibits the mRNA expression and production of MIP-1α from LPS-stimulated PMN at early time points, but augments MIP-1α expression later on.[28] The latter augmentation appears to be mediated by the autocrine effect of endogenously generated TNFα.[28] We attempted to elucidate the molecular basis of the regulation of LPS-induced MIP-1α gene expression in PMN, and of the early inhibitory action of IFNγ on this response. Nuclear run-on analyses revealed that LPS increases the transcriptional rate of the MIP-1α gene after 4 h of stimulation, and that IFNγ markedly inhibits the rate of MIP-1α gene transcription in LPS-activated neutrophils.[18] In agreement with the findings of Kasama et al,[28] IFNγ did not affect MIP-1α mRNA stability in PMN treated with LPS for a brief period (2-4 h).[18] This indicates that, at early time points, IFNγ does not inhibit the LPS-induced MIP-1α gene expression through post-transcriptional events, but only through an effect at the level of transcription.

Under some circumstances however, MIP-1α mRNA accumulation can be also modulated at the level of mRNA stability in neutrophils. For instance, after stimulation of neutrophils with LPS for 24 h, the stability of MIP-1α mRNA was shown to be significantly prolonged by IFNγ.[28] Whether this long-term effect is directly attributable to IFNγ, or perhaps mediated by endogenously produced TNFα,[28] was unfortunately not investigated. Another example of post-transcriptional regulation of MIP-1α mRNA is the effect of IL-10 towards MIP-1α gene expression.[7] In contrast to IFNγ, IL-10 was shown to inhibit the LPS-induced MIP-1α mRNA accumulation through enhanced mRNA degradation.[7] In this regard, we found that IL-10 does not inhibit the rate of LPS-stimulated MIP-1α gene transcription in PMN (our unpublished observations), therefore confirming that IL-10 primarily acts at the level of MIP-1α mRNA stability. Finally, mRNA stabilization analyses demonstrated that MIP-1α mRNA isolated from PMN stimulated in the presence of GM-CSF and LPS had a prolonged half-life, relative to LPS alone.[6] This helped to explain why PMN stimulated in the presence of both GM-CSF and LPS demonstrated an enhanced and prolonged expression for both MIP-1α mRNA and protein, as compared to LPS alone.[6] Conversely, GM-CSF alone failed to induce any expression of MIP-1α,[6] a finding later confirmed by Hachicha et al.[29] Although no studies on MIP-1β gene transcription have been reported to date, MIP-1β mRNA expression was influenced by IL-10 and IFNγ in the same manner as MIP-1α, that is, at the post-transcriptional level.[7,28] Stability of MIP-1β mRNA in LPS-treated neutrophils was significantly prolonged by IFNγ at 24 h,[28] but not at the 4 h time-point, and was markedly diminished by IL-10, at both 4 and 8 h time points.[7]

Another cytokine mRNA that in neutrophils appears to be mainly regulated at the post-transcriptional level is that encoding IL-1ra. For instance, the augmented expression of IL-1ra mRNA in PMN treated with IL-13[3] and TGFβ1[4] was shown to depend on a marked increase of IL-1ra transcript stability induced by both IL-13 and TGFβ$_1$. A possible transcriptional induction of IL-1ra gene could not be, however, excluded because ACT D partially blocked the enhancing effect of both IL-13[3] and TGFβ1.[4] In a previous study, we had also showed that release of IL-1ra from LPS-stimulated PMN was markedly potentiated in the presence of IL-10, and that this upregulation was associated with an enhanced stabilization of IL-1ra mRNA by IL-10.[25] The half-life of IL-1ra mRNA was prolonged in PMN stimulated in the presence of IL-10 and LPS, as compared with cells stimulated with LPS alone, whereas the half-life of IL-1β mRNA was unchanged[25] (Fig. 6.3). In spite of the fact that in the myelomonocytic cell line, THP-1, IL-10 also caused an upregulation of IL-1ra mRNA and production in LPS-treated cells,[30] no alteration by IL-10 of the IL-1ra mRNA half-life was observed.[30] These results imply that different regulatory mechanisms underlie the effects of IL-10 towards IL-1ra mRNA expression in PMN and THP-1 cells treated with LPS.[25,30] Consistent with our findings, it was reported that the suppression by IL-10 of TNFα and IL-1 mRNA accumulation in LPS-stimulated murine macrophages depends on de novo protein synthesis, as opposed to transcriptional inhibition of TNFα or IL-1 genes.[31] Quite surprisingly, the AUUUA-rich motifs believed to be involved in the regulation of mRNA stability[15] are not present in IL-1ra mRNA.[32,33] This therefore suggests that IL-1ra mRNA stability may be regulated in a unique manner. It is for example possible that in neutrophils IL-10 reduces the expression of a nuclease which selectively degrades IL-1ra mRNA, or that it up-regulates the expression of a factor that decreases the susceptibility of IL-1ra mRNA to the action of such a nuclease. In this respect, Bogdan and colleagues postulated the possible existence of nucleolytic activities controlling TNFα and IL-1 mRNA expression, which IL-10 could modulate.[31] At this point however, other effects of IL-10 at the level of IL-1ra transcription, translation or secretion in PMN cannot be excluded.

TRANSLATIONAL REGULATION

A series of observations indicate that in neutrophils, mechanisms other than gene transcription or mRNA stabilization are involved in the regulation of cytokine production. For example, a clear demonstration of a form of translational control in PMN has been shown for IL-1. Lord et al[34] observed that while LPS induced the expression of IL-1α and IL-1β mRNA in PMN, the latter were much less efficient in translating these transcripts when compared with peripheral blood mononuclear cells (PBMC). Similarly, Jack and Fearon found that PMN were less efficient than monocytes in synthesizing the CR1 protein.[35] In contrast, both PMN- and PBMC-derived IL-1 mRNAs were translated with equal efficiency in an in vitro protein synthesis system.[34] Conversely, Malyak et al[36] reported that IL-4 markedly decreased the total IL-1β protein synthesis in LPS-induced PMN, without reducing LPS-stimulated IL-1β mRNA levels. Under the same conditions, the combination of IL-4 and LPS resulted in a substantial increase in the total synthesis of IL-1ra (relative to LPS alone), but in this case the steady-state IL-1ra mRNA levels paralleled protein levels.[36] These results again suggest that the effects of IL-4 were mediated at the translational level for IL-1β production.[36] Similarly, a moderate reduction in the steady-state level of IL-1β mRNA was observed in neutrophils and monocytes cultured with both IL-10 and LPS[27] (relative to LPS alone); despite these lower levels of IL-1β mRNA, there was virtually no detectable change in IL-1β protein translated from this mRNA, suggesting again a control by IL-10 at a translational level of IL-1β mRNA.[27]

A possible explanation for the relative inefficiency of neutrophils to translate of IL-1α and IL-1β mRNA, is that PMN have a low number of ribosomes,[37] which limits the rate of total protein synthesis. Alternatively, PMN might contain a specific inhibitor of translation of IL-1 mRNA, while monocytes do not. A further possibility is that LPS predisposes neutrophils, in a sort of priming process, to effectively translate IL-1 mRNA, as soon as a second stimulus interacts with the cell. In this regard however, Lord and colleagues[34] reported that PMN stimulated with LPS plus phorbol myristate acetate (PMA), or fMLP, or the calcium ionophore, A23187, did not display a significantly increased synthesis of IL-1α and IL-1β proteins, as compared to LPS alone.[34]

Control at the translational level has been also suggested for IL-1ra and MIP-1α mRNA under specific experimental conditions. For instance, while pretreatment of neutrophils with TNFα before stimulation with calcium pyrophosphate dihydrate (CPPD) microcrystals upregulates the mRNA levels of IL-1ra mRNA as compared with those induced by TNFα alone, cell-associated and secreted protein levels of IL-1ra were inhibited in the presence of CPPD.[38] Under the same conditions, CPPD crystals synergistically increased both the mRNA and protein levels of IL-1β in TNFα-treated neutrophils. These data suggest that inhibition of IL-1ra synthesis by CCPD in TNFα-treated neutrophils occurs at the translational level.[38] Similarly, both CPPD and monosodium urate monohydrate (MSU) inhibited the immunodetectable MIP-1α induced by TNFα, without affecting MIP-1α steady-state mRNA levels.[29] Since the crystals neither enhanced the degradation of MIP-1α protein, nor interfered with the immunodetection of MIP-1α by ELISA, the data suggested that the inhibitory effect of microcrystals towards MIP-1α protein production, is primarily translational.[29]

POST-TRANSLATIONAL REGULATION

A likely example of regulation of secretion is that of IL-1ra production by neutrophils. As already discussed in chapter 4, Malyak et al[36] reported that LPS, GM-CSF, IL-4, and TNFα individually stimulate the production of IL-1ra protein by cultured PMN. However, less than 50% of the synthesized IL-1ra protein was secreted under all culture conditions, except for the combination of LPS and IL-4, which was more effective. Therefore, while the synthesis of IL-1ra was increased under these various conditions, the majority of the protein was not secreted but remained cell-associated. Conversely, when PMN were stimulated with LPS and IL-4, a more efficient secretion of IL-1ra was observed.[36]

Recent experiments performed in my laboratory revealed that the extracellular production of GROα by PMN did not always correlate with equivalent changes at the level GROα mRNA expression, suggesting that production of GROα in neutrophils is regulated post-transcriptionally.[39] For instance, we found that culture of PMN with LPS or fMLP strongly promoted GROα mRNA accumulation, at levels which were approximately 3- to 4-fold higher than those obtained following phagocytosis of IgG-opsonized S.cerevisiae (Y-IgG).[39] However, culture supernatants from fMLP-stimulated neutrophils contained amounts of GROα that were not significantly higher than those from untreated cells. Assuming that GROα mRNA is normally translated, a possible explanation would be that GROα is released, but rapidly degraded by the proteolytic enzymes that are simultaneously secreted in response to fMLP. Conversely, it cannot be excluded that in spite of its ability to strongly induce GROα mRNA, fMLP fails to provide the intracellular signals necessary to either translate or secrete GROα. If so, this would suggest that GROα production is controlled at the translational or post-translational level, as previously observed for GROα during malignant transformation of normal melanocytes.[40] Clearly, measurements of intracellular GROα levels under these different conditions are required before any conclusion can be reached. Worthy of note is that while fMLP also induces also high levels of IL-1β mRNA (Fig. 6.1), the corresponding protein is undetectable in the cell-free supernatants (our unpublished observations). Whether GROα and IL-1β mRNA transcripts are subjected to the same regulatory mechanisms in fMLP-stimulated neutrophils, remains to be established. In contrast to fMLP, Y-IgG is an extremely potent inducer of the extracellular production of GROα by PMN, being approximately 2- to 3-fold more effective than LPS or TNFα.[39] However, Y-IgG is a very weak inducer of GROα mRNA. These findings indicate that Y-IgG-phagocytosis regulates GROα production at the level of secretion, and is in keeping with our previous observations, that Y-IgG-stimulated PMN secrete IL-8 more efficiently than LPS- or TNFα-activated cells.[26]

We recently provided clear experimental evidence that in neutrophils, IL-8 production can be regulated at the level of secretion.[26] To better understand whether the effects of IFNγ on agonist-induced IL-8 expression might reflect changes in IL-8 synthesis or secretion, we separately quantified the IL-8 immunoreactivity that remained cell-associated, and that was released by the cells. Those studies brought forward a number of interesting observations.[26] First, the total production of IL-8 in LPS- and Y-IgG-treated PMN (as well as in resting cells), continuously increased, up to 18 h. Second, the amounts of IL-8 released by resting PMN, as well as by LPS-, TNFα- and Y-IgG-activated cells, increased throughout the incubation period, but the amounts of secreted IL-8 were inferior to those that were cell-associated, except for Y-IgG. Third, although the accumulation of IL-8 mRNA paralleled the increase in total IL-8 protein levels in both LPS- and Y-IgG-activated PMN, the latter secreted IL-8 more efficiently than LPS-stimulated cells (see also Fig. 6.4). Fourth, in PMN pretreated with IFNγ and then stimulated for up to 6 h, the total synthesis of IL-8 was significantly lower than when the cells were exposed to the stimuli in the absence of IFNγ. However, total IL-8 production after 18 h of incubation with various stimuli, was not significantly affected by the presence or absence of IFNγ, except for Y-IgG-treated cells. All these effects were paralleled by changes at the mRNA level. Moreover, the percentage of IL-8 secreted after stimulation with LPS, and TNFα for up to 18 h, or with Y-IgG for 2 h, was significantly higher in PMN that were pretreated with IFNγ in comparison with those untreated. Thus, even though IFNγ-treated PMN synthesized less IL-8 than untreated PMN, they secreted IL-8 more efficiently after stimulation with LPS or TNFα, at all time points examined. Lastly, IL-8 accumulation was shown to be also under translational control, as revealed by the fact that CHX, but not ACT D, inhibited the accumulation of cell associated IL-8 after culture at 37°C for 2 h.[41]

Altogether, these studies indicated that the up-regulatory effect of IFNγ on LPS-, and TNFα-induced secretion of IL-8 observed after 18 h, were largely explained by the potentiating effect of IFNγ at the level of IL-8 secretion. Therefore, in addition to transcriptional[9] and post-transcriptional regulatory[7,20] mechanisms, the extracellular yield of IL-8 can be also controlled at the secretional level.[26]

ADDITIONAL REMARKS

I conclude this chapter by making some speculations on putative mechanisms controlling TNFα production in PMN. Based on studies performed mainly with murine macrophages,[42] TNFα synthesis is believed to be tightly regulated at many levels, including translational.

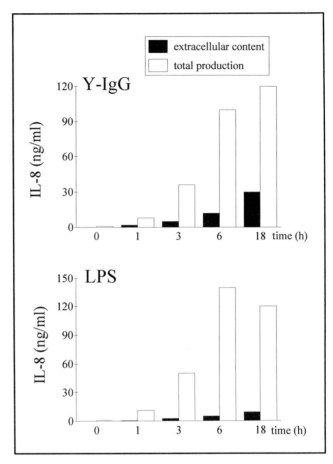

Fig. 6.4. Distribution of total IL-8 between cell-associated and cell-free supernatants in neutrophils stimulated with Y-IgG or LPS.

In this respect, Lindemman et al[43] reported that in human PMN, the induction by GM-CSF of TNFα mRNA accumulation is not accompanied by TNFα synthesis or release. Although the reasons for this lack of TNFα synthesis were not further investigated, it nevertheless appears that TNFα production might be subjected to a translational regulation of TNFα. It is also conceivable that by increasing the steady state level of TNFα mRNA, GM-CSF may act as a "priming" agent for a subsequent triggering stimulus, as observed for other PMN functions.[44] In another study, Van Dervort et al[45] demonstrated that in neutrophils, nitric oxide-generating compounds increased the LPS-induced TNFα production, yet without increasing TNFα mRNA levels. It must be stressed however, that Northern blots were performed at a single time point (1 h). Should Northern blot experiments performed at later time points confirm the lack of effect of nitric oxide-generating compounds towards TNFα mRNA steady state levels, then the enhancement of TNFα production by these compounds would likely reflect a modulation of translation or secretion.[45]

REFERENCES

1. Marucha PT, Zeff RA, Kreutzer DL. Cytokine-induced IL-1β gene expression in the human polymorphonuclear leukocyte: transcriptional and post-transcriptional regulation by tumor necrosis factor and IL-1. J Immunol 1991; 147:2603-2608.
2. Re F, Mengozzi M, Muzio M et al. Expression of interleukin 1 receptor antagonist by human circulating polymorphonuclear cells. Eur J Immunol 1993; 23:570-573.
3. Muzio M, Re F, Sironi M et al. Interleukin-13 induces the production of Interleukin-1 receptor antagonist (IL-1ra) and the expression of the mRNA for the intracellular (keratinocyte) form of IL-1ra in human myelomonocytic cells. Blood 1994; 83:1738-1743.
4. Muzio M, Sironi M, Polentarutti N et al. Induction by transforming growth factor-β1 of the interleukin-1 receptor antagonist and of its intracellular form in human polymorphonuclear cells. Eur J Immunol 1994; 24:3194-3198.
5. Wei S, Blanchard DK, Liu JH et al. Activation of tumor necrosis factor-α production from human neutrophils by IL-2 via IL-2Rβ. J Immunol 1993; 150:1979-1987.
6. Kasama T, Strieter RM, Standiford TJ et al. Expression and regulation of human neutrophil-derived macrophage inflammatory protein 1-alpha. J Exp Med 1993; 178:63-72.
7. Kasama T, Strieter RM, Lukacs NW et al. Regulation of neutrophil-derived chemokine expression by IL-10. J Immunol 1994; 152:3559-3569.
8. Strieter RM, Kasahara K, Allen RM et al. Cytokine- induced neutrophil-derived Interleukin-8. Am J Pathol 1992; 141:397-407.
9. Cassatella MA, Guasparri I, Ceska M et al. Interferon-γ inhibits interleukin-8 production by human polymorphonuclear leukocytes. Immunology 1993; 78:177-184.
10. Cassatella MA, Aste M, Calzetti F et al. Studies on the regulatory mechanisms of Interleukin-8 gene expression in resting and IFNγ-treated neutrophils. Evidence on the capability of staurosporine of inducing the production of IL-8 by human neutrophils. Biochem Biophys Res Commun 1993; 190:660-667 (published erratum in 1993; 192:324).
11. Cassatella MA, Gasperini S, Calzetti F et al. Lipopolysaccharide-induced interleukin-8 gene expression in human granulocytes: transcriptional inhibition by interferon-γ. Biochem J 1995; 310:751-755.
12. Cicco NA, Lindemann A, Content J et al. Inducible production of interleukin-6 by human neutrophils: role of Granulocyte-Macrophage Colony-Stimulating Factor and tumor necrosis factor alpha. Blood 1990; 75:2049-2052.
13. Wei S, Liu JH, Blanchard DK et al. Induction of IL-8 gene expression in human polymorphonuclear neutrophils by recombinant IL-2. J Immunol 1994; 152:3630-3636.

14. Carter BZ, Malter JS. Regulation of mRNA stability and its relevance to disease. Lab Invest 1991; 65:610-621.
15. Caput D, Beutler D, Hartog K et al. Identification of a common nucleotide sequence in the 3' untranslated region of mRNA molecules specifying inflammatory mediators. Proc Natl Acad Sci USA 1989; 83:1670-1674.
16. Brawerman G. Mechanisms of mRNA decay. TIBTECH 1990; 8:171-174.
17. Shaw G, Kamen R. A conserved AU sequence from the 3' untranslated region of GM-CSF mRNA mediates selective nRNA degradation. Cell 1986; 46:659-667.
18. Cassatella MA. Interferon-γ inhibits the lipopolysaccharide-induced macrophage inflammatory protein-1α gene transcription in human neutrophils. Immunol Letters 1996; in press.
19. Re F, Muzio M, De Rossi M et al. The type II receptor as a decoy target for IL-1 in polymorphonuclear leukocytes: characterization of induction by dexamethasone and ligand binding properties of the released receptor. J Exp Med 1994; 179:739-743.
20. Wang P, Wu P, Anthes JC et al. Interleukin-10 inhibits Interleukin-8 production in human neutrophils. Blood 1994; 83:2678-2683.
21. Cassatella MA, Bazzoni F, Calzetti F et al. Interferon-γ transcriptionally modulates the expression of the genes for the high affinity IgG Fc receptor (FcγR-I) and the 47 kDa cytosolic component of NADPH oxidase in human polymorphonuclear leukocytes. J Biol Chem 1991; 266: 22079-22082.
22. Cassatella MA, Hartman L, Perussia B et al. TNFa and immune interferon synergistically induce cytochrome b_{-245} heavy chain gene expression and NADPH oxidase in human leukemic myeloid cells. J Clin Invest 1989; 83:1570-1579.
23. Cassatella MA, Anegnon I, Cuturi MC et al. FcγR(CD16) interaction with ligand induces Ca^{2+} mobilization and phosphoinositide turnover in human natural killer cells. J Exp Med 1989; 169:549-567.
24. Marucha PT, Zeff RA, Kreutzer DL. Cytokine regulation of IL-1β gene expression in the human polymorphonuclear leukocyte. J Immunol 1990; 145:2932-2937.
25. Cassatella MA, Meda L, Gasperini S et al. Interleukin 10 up-regulates IL-1 receptor antagonist production from lipolysaccharide-stimulated human polymorphonuclear leukocytes by delaying mRNA degradation. J Exp Med 1994; 179:1695-1699.
26. Meda L, Gasperini S, Ceska M et al. Modulation of proinflammatory cytokine release from human polymorphonuclear leukocytes by gamma interferon. Cell Immunol 1994; 57:448-461.
27. Jenkins JK, Malyak M, Arend, WP. The effects of Interleukin-10 on Interleukin-1 receptor antagonist and Interleukin-1β production in human monocytes and neutrophils. Lymphokine Cyokine Res 1994; 13:47-54.
28. Kasama T, Strieter RM, Lukacs NW et al. Interferon gamma modulates the expression of neutrophil-derived chemokines. J Invest Med 1995; 43:58-67.

29. Hachicha M, Naccache PH, and McColl SR. Inflammatory microcrystal differentially regulate the secretion of macrophage Inflammatory protein 1 and interleukin 8 by human neutrophils: a possible mechanism of neutrophil recruitment to sites of inflammation in synovitis. J Exp Med 1995; 182:2019-2025.
30. Kline JN, Fisher PA, Monick MM et al. Regulation of Interleukin-1 receptor antagonist by Th1 and Th2 cytokines. Am J Physiol 1995; 269:L92-98.
31. Bogdan C, Paik J, Vodovotz Y et al. Contrasting mechanisms for suppression of macrophage cytokine release by transforming growth factor-β and interleukin-10. J Biol Chem 1992; 267:23301-23308.
32. Eisenberg SP, Evans RJ, Arend WP et al. Primary structure and functional expression from complementary DNA of a human interleukin 1 receptor antagonist. Nature 1990; 343:341-343.
33. Carter DB, Deibel MR, Dunn CJ et al. Purification, cloning, expression and biological characterization of an IL-1 receptor anatgonist protein. Nature 1990; 344:633-638.
34. Lord PCW, Wilmoth LMG, Mizel SB et al. Expression of interleukin-1 alpha and β genes by human blood polymorphonuclear leukocytes. J Clin Invest 1991; 87:1312-1321.
35. Jack RM, Fearon DT. Selective synthesis of mRNA and proteins by human blood neutrophils. J Immunol 1988; 140:4286-4293.
36. Malyak M, Smith MF, Abel AA et al. Peripheral blood neutrophil production of IL-1ra and IL-1β. J Clin Immunol 1994; 14:20-30.
37. Bainton DF, Ullyot JL, Farquhar MG. The development of neutrophilic polymorphonuclear leukocytes in human bone marrow. J Exp Med 1971; 134:907-934.
38. Roberge CJ, de Medicis R, Dayer JM et al. Crystal-induced neutrophil activation. V. Differential production of biologically active IL-1 and IL-1 receptor antagonist. J Immunol 1994; 152:5485-5494.
39. Gasperini S, Calzetti, F, Russo MP. De Gironcoli M. and Cassatella MA. Regulation of GROα production in human granulocytes. J Inflamm 1995; 45:143-151.
40. Bordoni R, Fine R, Murray D et al. Characterization of the role of melanoma growth factor stimulatory activity (MGSA) in the growth of normal melanocytes, nevocytes, and malignant melanocytes. J Cell Biochem 1990; 44:207-219.
41. Kuhns D, Gallin JI. Increased cell-associated IL-8 in human exudative and A23187-treated peripheral blood neutrophils. J Immunol 1995; 154:6556-6562.
42. Han J, Brown T, Beutler D. Endotoxin-responsive sequence control cachectin/tumor necrosis factor biosynthesis at the translational level. J Exp Med 1990; 171:465-475.
43. Lindemann A, Riedel D, Oster W et al. Granulocyte-Macrophage Colony-Stimulating Factor induces cytokine secretion by humam polymorphonuclear leukocytes. J Clin Invest 1989; 83:1308-1312.

44. Lopez AF, Williamson DJ, Gamble JR et al. Recombinant human granulocyte-macrophage colony-stimulating factor stimulates in vitro mature neutrophil and eosinophil function, surface receptor expression and survival. J Clin Invest 1986; 78:1220-1228.
45. Van Dervort AL, Yan L, Madara PJ et al. Nitric oxide regulates endotoxin-induced TNFα production by human neutrophils. J Immunol 1994; 152:4102-4109.

CHAPTER 7

PRODUCTION OF CYTOKINES BY NEUTROPHILS ISOLATED FROM INDIVIDUALS AFFECTED BY DIFFERENT HUMAN PATHOLOGIES

Studies that have examined the capacity of neutrophils to produce cytokines in human pathological conditions are not as many as those regarding mononuclear cells. However, considering that neutrophils are abundant in the circulation, and readily accessible to experimental investigation, I am sure that the number of such studies will rapidly grow. A concise description of what has been reported thus far is presented below.

PRODUCTION OF INTERLEUKIN-1

IL-1β ranks among the genes that are induced in neutrophils and monocytes/macrophages by endotoxin, and that appear to play an essential role in the pathogenesis of sepsis syndrome.[1] McCall et al[2] reported that peripheral blood (PB) polymorphonuclear leukocytes of patients with the sepsis syndrome (sepsis-PMN) were, in vitro, tolerant to endotoxin-induced expression of the IL-1β gene in vitro. The tolerance consisted in a combined reduction in lipopolysaccharide (LPS)-stimulated levels of IL-1β mRNA and a decreased synthesis of the immunoreactive IL-1β protein. McCall and colleagues also attempted to identify the mechanisms responsible for the tolerance of sepsis-PMN to endotoxin.[2] They excluded that it was due to the loss of the CD14 surface protein, the receptor required for endotoxin-mediated gene

---

Cytokines Produced by Polymorphonuclear Neutrophils: Molecular and Biological Aspects, by Marco A. Cassatella. © 1996 R.G. Landes Company.

induction in PMN,[3,4] or that it was the result of a global reduction in the functional responses of PMN.[2] The down-regulation of IL-1β gene in sepsis-PMN occurred concomitantly with an up-regulation of the constitutive expression of the type II IL-1 receptor (IL-1RII),[5] and did not persist in PMN of patients recovering from the sepsis syndrome. Tolerance involved specific signal transduction pathways triggered by endotoxin, since sepsis-PMN normally synthesized IL-1β in response to *S.aureus,* and secreted elastase.[2] Interestingly, tolerance was not limited to infection by Gram-negative bacteria, but was also observed when the sepsis syndrome was apparently induced by Gram-positive bacteria, Rickettsia, *Candida* species, or staphylococcal exotoxins, and not in patients seriously ill without detectable infection. The physiological significance of the tolerance to endotoxin for IL-1β gene expression in PMN, or for TNFα, IL-1β and IL-6 in blood monocytes,[6] may represent an attempt by the host to protect itself from the adverse effects of intravascular disseminated inflammation.

Cassone et al[7] assayed neutrophils isolated from patients at different stages of human immunodeficiency virus (HIV) infection, for IL-1β and IL-6 production and incubated PMN for 18 h with MP-F2, a mannoprotein from *C.albicans*. Neutrophils from HIV-infected patients were able to synthesize amounts of IL-1β and IL-6 comparable to those made by neutrophils of healthy subjects, suggesting that cells from HIV+ patients were good responders to activating signals. However, preliminary results obtained in my laboratory would instead suggest that, in HIV+ patients, the ability of PMN to produce specific cytokines in response to LPS is significantly altered. In fact, the production of IL-8 and IL-12p40 has been found to be lower than in HIV negative PMN. In contrast, the ability of IFNγ to modulate the production of the various cytokines seemed not to be modified in HIV+ patients. Interestingly, our data also provided evidence that in HIV+ patients, the observed dysregulations of PMN ability to produce specific cytokines are different than those observed in autologous peripheral blood mononuclear cells (PBMC).

NEUTROPHIL-DERIVED CYTOKINES IN RHEUMATOID ARTHRITIS

Rheumatoid arthritis (RA) is a systemic autoimmune disease characterized by chronic inflammation of the synovium, which often leads to the destruction of articular cartilage and juxta-articular bone. Although the etiology of RA is unknown, considerable evidence suggests that cytokines play a critical role in the pathogenesis of RA, particularly those belonging to the IL-1 system, even though they are not the only ones.[8] In view of the fact that neutrophils are present in the synovial fluid (SF) of patients with RA,[9] several groups have investigated whether these cells can be a potential source of cytokines in the SF. A study by Malyak et al,[10] for example, has suggested that neutrophils might significantly contribute to the total IL-1 receptor antagonist (IL-1ra) levels in SF in patients with active RA and other rheumatologic disorders, since they found a very strong correlation between SF IL-1ra levels and the number of neutrophils present in these fluids, whereas the correlation between SF IL-1ra levels and the number of mononuclear cells was not significant. Isolated SF PMN contained preexisting IL-1β and IL-1ra protein in the absence of mRNA, and both LPS and GM-CSF induced modest increases in IL-1β and IL-1ra mRNA and protein by cultured SF PMN, as well as by normal cultured PMN. Thus, with regard to the IL-1β and IL-1ra proteins, PMN isolated from inflammatory SF were found to be qualitatively similar to PMN from normal peripheral blood and not to be activated in vivo. Interestingly, SF samples from patients with noninflammatory arthropathies contained undetectable levels of IL-1ra, and therefore it can be inferred that normal SF does not contain IL-1ra.[10] The ability of RA blood- and SF-derived PMN to produce IL-1α and IL-1β, in the absence or presence of LPS, was also subsequently assessed by Dularay et al.[11] No production of IL-1α or IL-1β by SF-PMN was found. By contrast, blood PMN from 3/8 RA patients produced IL-1. These observations were substantially confirmed by Quaile et al[12], who reported that in RA patients, blood neutrophils, but not SF-PMN, contained significant levels of IL-1β mRNA. Their results imply that activation of IL-1β expression by PMN in RA occur in the circulation before the cells enter diseased joints. Recently, Beaulieu and McColl[13] found RA SF-neutrophils to be significantly less efficient in producing IL-1ra, compared with matched PB neutrophils. The spontaneous, GM-CSF- or TNFα-induced production of IL-1ra by SF-neutrophils were all significantly decreased when compared with PB neutrophils isolated from the same individuals. Under the same experimental conditions, production of both IL-1β and IL-8 was up-regulated, suggesting that there was no a general down-regulation of cell function in SF neutrophils. These results were also paralleled by a comparable modulation at the level of cytokine mRNA expression, as determined by Northern blot analysis.[13] Beaulieu and McColl therefore concluded that neutrophils are likely to be an important source of IL-8 and IL-1β in the RA joint, and that SF-neutrophils appear incapable of mounting a response as high as that of PB neutrophils in terms of IL-1ra production.[13] Furthermore, the potential ability of neutrophil-derived IL-8 to play an important role in attracting neutrophils to the SF, also suggested a possible role of these cells in the perpetuation of inflammation in RA. However, a novel finding in this area of research has been recently provided by Koch et al.[14] The latter authors found, in fact, significantly greater levels of antigenic GROα in SF from patients with RA as compared with osteoarthritis (OA) or other noninflammatory arthritides. Importantly, this GROα accounted for 28% of the chemotactic activity for PMN found in RA SF,[14] suggesting that GROα plays an important role in the migration of PMN into the inflamed RA joints. They also examined whether PMN (and other cells) obtained from RA SF generated GROα. They found that both RA SF PMN, as well as normal PB PMN, produced significant amounts of GROα, either constitutively or after stimulation with LPS.[14] Production of GROα and other chemokines by neutrophils and other cells may lead to the recruitment of more leukocytes to the joint, and therefore perpetuate RA.

PRODUCTION OF INTERLEUKIN-8

Some investigators have examined the IL-8 producing capacity of PMN in diseases other than RA.[13]

For example, Hsieh and coworkers[15] examined PMN from patients with systemic lupus erythematosus (SLE). PMN from isolated SLE patients exhibit several functional abnormalities,[16] and, in fact, an increased susceptibility to infections is one of the hallmark of SLE.[17] Interestingly, Hsieh et al[15] found that the spontaneous and LPS-stimulated production of IL-8 by the peripheral blood PMN of active SLE patients were impaired as compared to inactive SLE or healthy individuals. This impaired IL-8 production by SLE-PMN was not linked to the administration of steroid, because incubation of normal PMN or inactive SLE-PMN with prednisolone for 24 h did not affect IL-8 production.[15] However, the possibility that a long-term dose of immunosuppressive treatment may lead to a defective IL-8 production in active SLE patients could not be excluded. The results of Hsieh et al[15] suggest that a decreased IL-8 production is one of the functional defects of PMN in patients with active SLE, that might predispose to infection.

In another work,[18] Lin and Huang evaluated the gene expression and release of IL-8 by peritoneal macrophages (PM) and PMN during peritonitis caused by *S.aureus*, in uremic patients on continuous ambulatory peritoneal dialysis (CAPD). Their previous study indicated in fact that IL-8 was detectable in drain dialysate of uremic patients on CAPD during the early acute stage of peritonitis, but at variable levels, depending on the microorganism.[19] In their subsequent study,[18] Lin and Huang revealed that the IL-8 levels were highly correlated with the PMN count found in the drain dyalisate, and that the amount of IL-8 mRNA expression was also highly correlated with the PMN count, since both were high at the onset of peritonitis and then decreased together progressively.[18] However PM, expressed more IL-8 mRNA than PMN. Their data strongly suggest that through the release of IL-8, PMN may be considered potential contributors to the pathogenesis of peritoneal injury.

In a series of papers, Kuhns et al[20,21] observed that during the evolution of the inflammatory response associated with skin lesions raised by suction, IL-8 reached levels up to 175 ng/ml in the media bathing the lesions. Accumulation of IL-8 strictly correlated with the accumulation of the exudative neutrophils at this inflammatory site.[21] Furthermore, these neutrophils, exhibited 100-fold greater levels of cell-associated IL-8, and spontaneously released up to 50-fold more IL-8 than freshly isolated peripheral blood neutrophils from the same donors.[21] These data indicate that neutrophils that have undergone diapedesis and migrated to a inflammatory focus, have upregulated their production of IL-8, and suggest that, by releasing IL-8, PMN play a role in the autocrine regulation of the inflammatory response.

PRODUCTION OF OTHER CYTOKINES

Takeichi and colleagues[22] examined whether PMN derived from chronic inflammation can express cytokine genes in vivo. For this purpose, they focused on adult periodontitis, a chronic infectious disease associated with active tissue damage. PMN constitute the great majority (>95%) of cells in gingival crevicular fluid (GCF) obtained from gingival inflammation. Furthermore, in GCF, it is possible to reveal, with considerable frequency, significant levels of both IL-1 or TNFα.[23] Takeichi et al[22] found significant levels of biologically active IL-1α and IL-1β, but not TNFα or IL-6, in GCF. Very elegantly, by using reverse-transcriptase polymerase chain reaction (RT-PCR), associated with a slot blot analysis, they provided clear evidence that highly purified PMN (>99.5%) collected from GCF express IL-1α, IL-1β and TNFα mRNA, but not IL-6 transcripts. Furthermore, they showed that PB monocytes or PB lymphocytes strongly expressed IL-1α, IL-1β, TNFα, and IL-6 mRNA after stimulation in culture, whereas PMN, again, expressed IL-1α, IL-1β and TNFα transcripts, but not IL-6 messages.[22] These results, other than demonstrating the possibility that PMN migrated in inflamed tissue can produce proinflammatory cytokines which may have a role in the initiation and development of periodontal disease, further support the notion that PMN do not express or produce IL-6. In a further study, Tonetti et al[24] performed in situ hybridization of IL-8 and MCP-1 genes in frozen tissue sections from patients affected by periodontal infections. Maximal IL-8 expression was found in the junctional epithelium adjacent to the infecting microorganisms, where PMN infiltration was more prominent, whereas MCP-1 was expressed in the chronic inflammatory infiltrate and along the basal layer of the oral epithelium where only cells of the monocyte/macrophage lineage were present.[24] These topographically specific tissue locations of chemokine mRNA expression were consistent with a hypothetical establishment of a discrete chemotactic source for an effective local host defense.

In another study, Raqib et al[25] examined by immunochemistry cryopreserved tissues from *Shigella*-infected patients, to determine, at the single-cell level, the cytokine-producing cells during the early and late stages of shigellosis. *Shigella* infection is usually accompanied by an intestinal activation of epithelial cells, T cells, and macrophages within the inflamed colonic mucosa. Histopathologically, *Shigella* infection is characterized by the presence of chronic inflammatory cells with or without neutrophils and microulcers in the lamina propria, crypt distortion, and, less frequently, crypt abscess. Raqib and coworkers found that *Shigella*-infected patients had significantly higher numbers of cytokine producing cells for all the cytokines studied (IL-1α, IL-1β, IL-1ra, IL-4, IL-6, IL-8, IL-10, IFNγ, TNFα, TNFβ, TGFβ1, TGFβ2, TGFβ3), than the healthy controls. However, production of the various cytokines in the rectal biopsies during acute and convalescent periods was not significantly different between the two periods, with the exceptions of TGFβ and IL-1ra.[25] Interestingly, in the acute *Shigella* infection, PMN present in the crypt abscess were clearly seen to contain IL-1β.[25] This observation is relevant in view of current evidence obtained by Sansonetti and coworkers,[26] that IL-1 is a key player in the cascade mediating invasion and inflammation of the intestinal mucosa.

Very recently, Beil and colleagues[27] used an ultrastructural immunogold morphologic and morphometric analysis to identify the cellular and subcellular sites of TNFα in vivo, in colonic biopsies obtained from patients with Crohn's disease (CD). CD is a chronic inflammatory disorder of bowel of unknown etiology, in which there is increased evidence that TNFα is implicated in its pathophysiology.[28] In contrast to ulcerative colitis, which is another type of inflammatory bowel disease, neutrophils are not a frequent infiltrating cell type in bowel tissues of CD.[29] However, Beil's study identified TNFα expressed in tissue neutrophils, in addition to eosinophils, macrophages, mast cells, fibroblasts, epithelial and Paneth cells. In neutrophils, TNFα was not present in cytoplasmic granules, but rather, it was associated with the membranes of Golgi structures and cytoplasmic vesicles, or in lipid bodies.[27] Thus, while numbers of neutrophils are small in colonic CD samples, when present they contain TNFα.

Finally, Bortolami and colleagues assessed whether IFNα, normally used in the treatment of chronic viral hepatitis, could affect PMN-derived TNFα production.[30] They evaluated the effect of different doses of IFNα on TNFα production by resting and LPS-activated human neutrophils from normal and hepatitis C virus (HCV)-infected patients. Their results revealed that none of the IFNα concentrations (25-5000 U/ml) alone induced TNFα from PMN, and that TNFα production by PMN after LPS stimulation was similar in normal and HCV-infected patients. However, various doses of IFNα associated with LPS induced a marked increase in TNFα secretion by PMN from HCV-infected patients, but had minimal effects on healthy control PMN.[30] These data attributed a new property to the numerous biological effects of IFNα, which may be taken into account during IFNα therapy for HCV-related chronic hepatitis.

REFERENCES

1. Dinarello CA. The interleukin-1 family: 10 years of discovery. FASEB J 1994; 8:1314-1324.
2. McCall CE, Grosso-Wilmoth LM, LaRue K et al. Tolerance to endotoxin-induced expression of the Interleukin-1β gene in blood neutrophils of humans with the sepsis syndrome. J Clin Invest 1993; 91:853-861.
3. Haziot A, Tsuberi BZ, Goyert SM. Neutrophil CD14: biochemical properties and role in the secretion of tumor necrosis factor-α in response to lipopolysaccharide. J Immunol 1993; 150:5556-5565.
4. Wright SD, Ramos RA, Tobias PS et al. CD14, a receptor for complexes of lipoplysaccharide (LPS) and LPS binding protein. Science 1990; 249:1431-1436.
5. Fasano MB, Cousart S, Neal S et al. Increased expression of the interleukin-1 receptor on blood neutrophils of humans with the sepsis syndrome. J Clin Invest 1991; 88:1452-1459.
6. Munoz C, Carlet J, Fitting C et al. Dysregulation of an in vitro production by monocytes during sepsis. J Clin Invest 1991; 88:1747-1754.
7. Cassone A, Palma C, Djeu JY et al. Anticandidal activity and interleukin-1β and interleukin 6 production by polymorphonuclear leukocytes are preserved in subjects with AIDS. J Clin Mirobiol 1993; 31:1354-1357.
8. Arend WP, Dayer JM. Cytokines and cytokine inhibitors or antagonists in rheumatoid arthritis. Arthritis Rheum 1990; 33:305-315.
9. Firenstein GS, Zvaifler NJ. Rheumatoid arthritis. A disease of disordered immunity. In: Gallin JI, Goldstein IM, Snyderman R, eds. Inflammation: Basic Principles and Clinical Correlates. 2nd edition. New York: Raven Press, 1992; 959-975.
10. Malyak M, Swaney RE, Arend WP. Levels of synovial fluid Interleukin-1 receptor antagonist in rheumatoid arthritis and other arthropathies. Potential contribution from synovial fluid neutrophils. Arthritis Rheum 1993; 36:781-789.
11. Dularay B, Westacott CI, Elson CJ. IL-1 secreting cell assay and its application to cells from patients with rheumatoid arthritis. Br J Rheum 1992; 31:19-24.
12. Quayle JA, Adams S, Bucknall RC et al. Interleukin-1 expression by neutrophils in rheumatoid arthritis. Ann Rheum Dis 1995; 54:930-933.
13. Beaulieu AD, McColl S. Differential expression of two major cytokines produced by neutrophils, interleukin-8 and interleukin-1 receptor antagonist, in neutrophils isolated from the synovial fluid and peripheral blood of patients with rheumatoid arthritis. Arthritis Rheum 1994; 37:855-859.
14. Koch AE, Kunkel SL, Shah MR et al. Growth-related gene product α. A chemostatcic cytokine for neutrophils in rheumatoid arthritis. J Clin Invest 1995; 155:3660-3666.
15. Hsieh SC, Tsai CY, Sun KH et al. Decreased spontaneous and lipopolysaccharide stimulated production of interleukin-8 by polymorphonuclear neutrophils of patients with active systemic lupus erythematosus. Clin Exp

Rheum 1994; 12:627-633.
16. Landry M. Phagocyte function and cell-mediated immunity in systemic lupus erythematosus. Arch Dermatol 1977; 113:147-154.
17. Staples PJ, Gerding DN, Decker JL et al. Incidence of infection in systemic lupus erythematosus. Arthritis Rheum 1974; 17:1-10.
18. Lin CY, Huang TP. Gene expression and release of Interleukin-8 by peritoneal macrophages and polymorphonuclear leukocytes during peritonitis in uremic patients on continuous ambulatory peritoneal dialysis. Nephron 1994; 68:437-441.
19. Lin CY, Lin CC, Huang TP. Serial changes of IL-6 and IL-8 levels in drain dyalisate of uremic patients with continuous ambulatory peritoneal dialysis during peritonitis. Nephron 1993; 63:404-408.
20. Kuhns D, DeCarlo E, Hawk DM et al. Dynamics of the cellular and humoral components of the inflammatory responses elicited in skin blisters in humans. J Clin Invest 1992; 89:1734-1740.
21. Kuhns D, Gallin JI. Increased cell-associated IL-8 in human exudative and A23187-treated peripheral blood neutrophils. J Immunol 1995; 154:6556-6562.
22. Takeichi O, Saito I, Tsurumachi T et al. Human polymorphonuclear leukocytes derived from chronically inflamed tissue express inflammatory cytokines in vivo. Cell Immunol 1995; 156;296-309.
23. Kabashima H, Maeda K, Iwamoto Y et al. Partial characterization of an interleukin-like factor in human gingival crevicular fluid from patients with chronic inflammatory periodontal disease. Infect Immun 1990; 58:2621-2627.
24. Tonetti MS, Imboden MA, Gerber L et al. Localized expression of mRNA for phagocyte-specific chemotactic cytokines in human periodontal infections. Infect Immun 1994; 62:4005-4014.
25. Raqib R, Lindberg AA, Wretlind B et al. Persistence of local cytokine production in shigellosis in acute and convalescent stages. Infect Immun 1995; 63:289-296.
26. Sansonetti PJ, Arondel J, Cavaillon JM et al. Role of Interleukin-1 in the pathogenesis of experimental shigellosis. J Clin Invest 1995; 96:884-892.
27. Beil WJ, Weller PF, Peppercorn MA et al. Ultrastructural immunogold localization of subcellular sites of TNFα in colonic Crohn's disease. J Leuk Biol 58; 284-298:1995.
28. Derkx B, Taminiau J, Radema S et al. Tumor necrosis factor antibody treatment in Crohn's disease. Lancet 1994; 342:173-174.
29. Dvorak AM. Ultrastructural pathology of Crohn's disease. In: Goebell H, Peskar BM, Malchow H, eds. Inflammatory Bowel Diseases-Basic Research and Clinical Implications. Lancaster: MTP Press, 1988:3-41.
30. Bortolami M, Carlotto C, Fregona I et al. Effects of interferon-α on the production of tumor necrosis factor-α by polymorphonuclear cells. Fund Clin Immunol 1995; 3:153-156.

CHAPTER 8

CYTOKINE PRODUCTION BY NEUTROPHILS IN VIVO

In agreement with numerous in vitro observations, many in vivo studies have confirmed the possibility that polymorphonuclear leukocytes (PMN) might be a significant source of cytokines in inflammatory lesions. As mentioned in chapter 3 already, most of the latter studies have evaluated the production of cytokines by neutrophils in animals injected with lipopolysaccharide (LPS) under different conditions (Fig. 3.8). However, as we shall see, other experimental in vivo conditions provoke neutrophils to express cytokines, and in some of these situations, the production of neutrophil-derived cytokines appears to be fundamental for the evolution and/or resolution of the induced pathological process.[1] Table 8.1 summarizes all the results obtained to date regarding the production of cytokines by neutrophils in vivo, in many different experimental animal models.

EFFECT OF LPS ADMINISTRATION IN VIVO ON NEUTROPHIL-DERIVED CYTOKINES

To study neutrophil cytokine expression in vivo, the most widely used experimental animal system is an LPS-induced acute inflammation. One such example is the intratracheal injection (it) of endotoxin in the rat, which causes a dramatic influx of PMN into the bronchoalveolar space. For instance, when the kinetics of IL-1α/β and IL-1ra mRNA expression in the lung were investigated by Northern analysis after intravenous (iv) injection of 100 μg LPS, it was found that IL-1α/β mRNA expression peaked at 1 h, whereas IL-1ra peaked at 2 to 4 h, consistent with the hypothesis that IL-1ra acts to turn off the effects of IL-1 through a negative feedback mechanism.[2] In the same study, it injection of LPS had strikingly different consequences: IL-1α mRNA peaked at 2 to 6 h, whereas IL-1β/IL-1ra peaked at 6 h, and that was concurrent with the maximum influx of neutrophils.

Cytokines Produced by Polymorphonuclear Neutrophils: Molecular and Biological Aspects, by Marco A. Cassatella. © 1996 R.G. Landes Company.

Table 8.1. Cytokines produced by neutrophils in vivo

Experimental Animal Model	Cytokine(s) Produced	Source	References
New bone formation in rats	TGFβ	developing bone	53
New Zealand white rabbits injected intraperitoneally with casein	IL-1β	peritoneal exudate	14
Swiss albino outbred mice challenged by intraperitoneal injection of LPS	IL-6	peripheral blood	8
Lewis rats intratracheally injected with LPS	IL-1α/β and IL-1ra	bronchoalveolar lavage	2
CD-1 rats intratracheally instilled with LPS	MIP-2 and KC	bronchoalveolar lavage	5
Rats intraperitoneally injected with thioglycollate	MIP-2 and KC	peritoneal exudate	5
Immune complex peritonitis of WBB6F$_1$ mice	TNFα	peritoneal exudate	27
Sprague-Dawley rats intratracheally instilled with LPS	TNFα	bronchoalveolar lavage	3
Sprague-Dawley rats intravenously injected with LPS	IL-1β	pulmonary vasculature lung	6
BALB/c mice injected with a colon adenocarcinoma releasing G-CSF	IL-1α/β and TNFα	tumor-infiltrating cells	51
Sprague-Dawley rats intratracheally instilled with LPS	TNFα, IL-1β, IL-6, MIP-2	bronchoalveolar lavage	4
New Zealand white rabbits injected intraperitoneally with casein	IL-1β, IL-8 and MIP-1β	peritoneal exudate	13
Dog trachea superfused with supernatans from *Pseudomonas*	IL-8	recruited cells	34
Wistar rats instilled with bleomycin	MCP-1	bronchoalveolar lavage	22
Sprague-Dawley rats intratracheally inoculated with LPS	TNFα	bronchoalveolar lavage	12
Sheep infused with LPS	TNFα	pulmonary vasculature lung	7
BALB/c mice orogastrically infected with *Y. enterocolitica*	IL-1ra	blood, spleen, Peyer patches	32
Reperfusion of ischemic myocardium of New Zealand white rabbits	IL-8	post-ischemic myocardial tissue	16
Retinal ischemia and reperfusion of Sprague-Dawley rats	IL-1β	retinal tissue	20
New Zealand white rabbits exposed to hyperoxia	IL-8	bronchoalveolar lavage	26
New Zealand white rabbits intra-articularly injected with IL-8	IL-1β and IL-1ra	infiltrating leukocytes	31
C3H/HeN and C3H/HeJ mice intraperitoneally injected with LPS	TNFα/IL-1α, IL-10	lung tissue	11
Mice injected with *C. albicans*	IL-10 and IL-12	infiltrating leukocytes	1

Fractionation of alveolar macrophage-enriched and PMN-enriched subpopulations from the bronchoalveolar lavage (BAL) cells obtained after it injection of LPS revealed that neutrophils were the predominant source for both IL-1α/β and IL-1ra mRNA.[2] This study therefore uncovered that differences in the kinetics of IL-1 and IL-1ra mRNA expression in whole lung RNA preparations after iv or it injections of LPS were due to the contribution of PMN, which appeared in the lung in a large numbers, following it injections.[2] It can be hypothesized that the production of IL-1 by PMN plays a role in the activation of lymphocytes during the transition between acute neutrophilic and chronic mononuclear inflammation, whereas the synthesis of IL-1ra might function as a negative feedback mechanism by which neutrophils downregulate their own influx into inflammatory sites.

Using a similar rat model, Xing et al[3] initially investigated the time-dependent expression and potential cellular source of TNFα in the lung. By Northern blot analysis, both alveolar macrophages (AM) and PMN were found to express TNFα mRNA, but PMN displayed several times more TNFα mRNA than AM at 6 and 12 h after it instillation of LPS (25 μg/kg), in parallel with the peak of PMN infiltration. By in situ hybridization, most of the cells positive for TNFα mRNA seemed to be PMN localized within the inflamed tissue near bronchioles or vessels. By immunohistochemistry, TNFα protein was localized mainly to AM at early times after LPS challenge (1 to 3 h), whereas thereafter (6-12 h), PMN were the predominant source of TNFα protein. This work provided the first in vivo evidence that PMN can represent a significant source of TNFα at sites of acute inflammation. In a subsequent, more extensive study, Xing et al[4] demonstrated that LPS triggers a distinct cytokine response in the lung, by selectively increasing mRNA transcripts encoding TNFα, IL-1β, IL-6, and MIP-2 (which is functionally equivalent to IL-8), but not RANTES or TGFβ1, and that the cellular sources for these cytokines included AM and PMN. At a time (1 h) when only a minimal PMN infiltration was present, AM appeared to be the predominant source of all cytokines examined, whereas at later times (6 and 12 h) when PMN infiltration became maximal (88% PMN), PMN were the prominent source of these cytokines. A low, basal, noninducible signal for TGFβ1 (but not for RANTES) mRNA was detected in both AM and PMN. Interestingly, in situ hybridization of the lung tissue, revealed that amongst the cells which stained for MIP-2 mRNA in response to LPS, were, particularly, the PMN located in the vicinity of bronchioles and vasculature, but not within the vasculature itself.[4] That recruited PMN under similar experimental conditions (it instillation of 10 mg/kg LPS) could express MIP-2 and KC (the latter being the murine analogue of GROα/MGSA) had also been indirectly proved by Huang et al.[5] They showed that the expression of MIP-2 and KC by BAL cells was rapidly induced (30 min) and persisted for 16 h. Although their Northern analyses were performed on pooled BAL cell RNA, they found that following LPS instillation, these BAL cell population changed from predominantly (>95%) macrophages to mostly PMN (60% within 2 h, 91% after 16 h) after LPS instillation.[5] Furthermore, Huang et al[5] also mentioned that they found expression of MIP-2 and KC mRNA within exudative neutrophils obtained after intraperitoneal (ip) injection of thioglycollate into rats. Collectively these observations support the notion that in this lung model of LPS-elicited inflammation, AM appear to be the predominant sources of the cytokines examined when only a minimal PMN infiltration is present (1 h), whereas at later time points (6 to 16 h), infiltrating PMN become a significant source of cytokines.

Whereas all of the above data show that PMN located within the lung tissue can be a potent source of cytokines upon encounter with airway-derived LPS, a number of studies have shown that, conversely, circulating PMN can also constitute a prominent source of cytokines. Williams et al[6] showed that following iv infusion of LPS (3 mg/kg) for 2 h in rats, PMN rather than mononuclear cells (MNC) in the pulmonary vasculature were the major source of IL-1β transcripts detected in total vascular leukocytes. In contrast, no induction of IL-1β expression was observed in airway leukocytes or circulating leukocytes. Under these experimental conditions, LPS increased pulmonary vascular sequestration of leukocytes, recruiting most prominently an activated pool of neutrophils that were more adherent, primed for increased reactive oxygen intermediates (ROI) production, and that expressed increased IL-1β messages. Thus, this early study suggested a more prominent role than previously appreciated for sequestered neutrophils in sepsis-induced lung inflammation.[6] More recently, a study by Cirelli et al[7] confirmed that an accumulation of intravascular mononuclear phagocytes and neutrophils in the pulmonary circulation is also observed during a continuous infusion of endotoxin in sheep. These authors detected an increased cytoplasmic TNFα immunoreactivity in both mononuclear phagocytes and neutrophils sequestered in pulmonary arterioles, capillaries, and venules. Coincidentally, plasma levels of TNFα significantly increased, suggesting that both neutrophils and mononuclear phagocytes contributed to the rise in the circulating levels of TNFα, and the development of acute lung injury.[7]

Another model of LPS-induced inflammation was used by Terebuth et al.[8] These investigators performed immunohistochemical studies to localize cells expressing IL-6, in selected organs of normal and endotoxin-challenged NIH-Swiss outbred mice. In normal mice, a constitutive cytoplasmic IL-6 immunoreactivity was detected in blood monocytes and their precursors, in bone marrow and splenic stromal macrophages, and in granulocytes as well.[8] Interestingly, the authors noticed that certain methods of mechanical PMN enrichment could lead to an inability to detect IL-6 in PMN. As mentioned in chapter 4 with regard to the controversial issue of IL-6 production by human PMN, Melani et al[9] also reported that the handling procedures used to purify neutrophils seemed to specifically affect the constitutive and stimulated expression of IL-6. Terebuth et al[8] also observed that while significant serum levels of IL-6 were absent, cell-associated IL-6 bioactivity was found in circulating PMN but not in lymphocytes. However, after IP injection of LPS (40 mg/kg), there was a 2- to 3-fold increase in PMN cell-associated IL-6 bioactivity from 1 to 3 h, followed by an almost complete depletion at 6 h, suggesting that there was increased synthesis in PMN and that this population was a potential source of serum IL-6. Indeed, serum levels of IL-6 peaked at 3 h after LPS challenge and dropped significantly by 6 h. Interestingly, constitutive and increased intracellular IL-6 in circulating PMN was detected in the absence of IL-6 mRNA, which was instead present in granulocytic/monocytic progenitors in the bone marrow; in the latter cells, IL-6 transcripts increased with a similar time course after LPS challenge.[8] These data suggest a scenario in which circulating granulocytes bear IL-6 as a stored component, likely acquired during bone marrow maturation. During experimentally induced sepsis, PMN might release IL-6 as a result of appropriate signals received, for example, during margination or chemotaxis. Previously, the same authors, employed an identical model to demonstrate that the expression of IL-1 and TNFα (at times corresponding to elevated levels of these cytokines in the serum) was most prominent in Kupffer cells.[10] In contrast, IL-6 was not consistently expressed by Kupffer cells.[8] Together, their results also suggest a different pattern of expression (and likely of regulation) for IL-6, than for IL-1 and TNFα.

More recently, Nill et al[11] compared the temporal sequence of endotoxin-induced TNFα, IL-1α, and IL-10 gene expression and cellular localization of cytokine proteins in pulmonary tissue of two strains of mice that have a genetically based differential sensitivity to endotoxin. Cytokines were studied by quantitative polymerase chain reaction (PCR) and in situ hybridization in lung tissue harvested from endotoxin-sensitive C3H/HeN and endotoxin-resistant C3H/HeJ mice at different times after ip injection of LPS.[11] Although levels of TNFα mRNA and protein in the two mouse strains were similar at 1-2 h, the IL-1α gene and protein expression in pulmonary tissue isolated from endotoxin-resistant mice was lower at any time point examined.[11] But the most dramatic difference was found in the case of IL-10 mRNA and protein levels, which were upregulated and continued to increase over a 12 h time period in C3H/HeN mice, whereas they were basically undetectable in CH3/HeJ endotoxin-resistant mice. In both types of mouse strains, TNFα, IL-1α, and IL-10 immunoreactive proteins were localized primarily to the infiltrating neutrophils, as well as to alveolar macrophages and type II pneumocytes.[11] However, quantitation of neutrophil infiltration into pulmonary tissue demonstrated that there was a significantly decreased inflammatory infiltrate in pulmonary tissue isolated from CH3/HeJ mice following LPS-administration, which correlated with decreased levels of immunoreactive cytokine proteins within pulmonary cells.[11] The decreased neutrophil infiltration observed in pulmonary tissue of CH3/HeJ mice might have been due to the comparative differences in the proinflammatory cytokine environment, which might have lead to a decreased expression of adhesion molecules on vascular endothelial cells and leukocytes, or to a insufficient production of chemokines or chemotatctic factors, ultimately resulting in a diminished recruitment of leukocytes. The results of Nill et al[11] unequivocally implicate that infiltrating neutrophils are important cellular mediators of pulmonary tissue damage induced by endotoxin in the CH3H/HeN endotoxin-sensitive mice.

Finally, in a study recently performed to investigate the in vivo effects of ethanol on LPS-induced TNFα and nitric oxide synthase (iNOS) expression in the lung, ethanol intoxication was found to not affect TNFα mRNA expression in alveolar macrophages or recruited neutrophils, but to inhibit that of iNOS.[12]

OTHER IN VIVO MODELS OF ACUTE INFLAMMATION INVOLVING NEUTROPHILS

Mori et al[13] examined the RNA synthesis and related dynamic changes in neutrophils, during the course of acute inflammation in rabbits. They investigated, by Northern blot analysis, the expression of 12 different genes in peritoneal exudate neutrophils, harvested at 5 and 24 h after IP injection of casein.[13] The genes for IL-1α, TNFα and MCP-1 were below the detection levels during the entire inflammatory period of observation, suggesting that under such conditions, the expression of these genes is very low. By contrast, on the basis of the expression kinetics of the remaining nine genes, the authors classified them into three categories; noteworthy is that subdivision also coincided well with the functional aspect of the products of the expressed genes. The first group included γ-actin, MRP-8 and MRP-14, the latter two being calcium-binding proteins and components of a complex molecule with inhibitory activity against casein kinase I and II. These messages were constitutively expressed in blood neutrophils and were also rapidly induced after emigration into inflammatory sites. The second group of genes included IL-1β, IL-8, MIP-1β and the formyl-methionyl-leucyl-phenylalanine receptor (fMLP-R), which were induced rapidly after the onset of inflammation (2-5 h), but had returned to basal levels of expression by 24 h. In the case of IL-1β expression, this was in agreement with previous findings reported by the same group using immunocytochemical studies at a single-cell level,[14] which had shown that neutrophils synthesize IL-1β de novo at acute inflammatory sites in the peritoneal cavity of rabbits, but not in the peripheral blood.[14,15] IL-1β was in fact apparently produced by neutrophils exuded only at an early stage (within 8 h) of inflammation, but not at later periods, even though the cells seemed perfectly viable.[13,14] Interestingly, at no time point were macrophages the major IL-1β producer in this particular type of inflammation.[14] Therefore, it is supposed that the cell type contributing to IL-1 production in vivo is dependent upon the type of inflammation; production of IL-1β in the casein-induced peritonitis rabbit model depends mainly upon its biosynthesis by PMN. The functions of the products of the genes in the second group relate especially with chemotaxis, one of the hallmarks of early inflammation. To the third group of neutrophil genes expressed, only ferritin-related mRNAs (F and H chains) were ascribed, because they were induced slowly (4-7 h), and increased with the progression of the inflammatory process. This finding suggests in any case that neutrophils retain the potential for biosynthesis at later stages of inflammation. The study of Mori et al[13] not only underlined that neutrophils contribute to the acute inflammatory reactions by synthesizing a variety of proteins for a fairly long period, but also evidenced that this neutrophil response is regulated and subjected to a programmed sequence.

In another recent paper aiming to identify the neutrophil chemoattractants generated in a model of myocardial infarction in the rabbit, Ivey et al[16] attributed important roles to the complement fragment C5a, and to the chemokine IL-8. Ischemia induces all the typical changes characteristic of an acute inflammatory response, among which an early neutrophil accumulation is a prominent feature. A determinant step in neutrophil accumulation is the local generation of chemical signals responsible for leukocyte recruitment. Neutrophil accumulation is markedly accelerated during reperfusion after ischemia, and early studies have implicated PMN in the generation of tissue damage associated to reperfusion.[17,18] In their study, Ivey et al[16] demonstrated that immunoreactive C5a and IL-8 were present in myocardial tissue after ischemia and reperfusion, but the time course of their appearance were quite different. C5a was detected already after 5 min of the initiation of reperfusion, while IL-8 concentrations rose slowly and were significantly elevated at 1.5 h and highest at 4.5 h, in close parallel with leukocyte infiltration. Further experiments revealed that neutrophil depletion virtually abolished IL-8 generation in the myocardium, but had no influence on C5a generation. These results therefore shed light on some of the basic mechanisms involved in the process of neutrophil accumulation in myocardial tissue after ischemia and reperfusion. C5a is liberated from preformed substrates as early as a few minutes after the initiation of reperfusion and induces a first phase of neutrophil infiltration. Once in the tissue, neutrophils become the source of IL-8 in the myocardium, and this IL-8 generation is responsible for a subsequent neutrophil accumulation over the whole time period investigated.[16] Although drugs inhibiting arachidonic acid metabolism via the 5'-lipoxygenase pathway are able to suppress neutrophil accumulation and limit infarct size,[19] the study of Ivey et al[16] suggests that suppression of C5a and IL-8 generation might be expected to be an effective means of inhibiting neutrophil accumulation.

In a rat model of transient retinal ischemia, a condition which leads to neuronal damage, Hangai et al[20] studied the levels of IL-1 gene expression by semi-quantitative PCR, and also used in situ hybridization histochemistry for cellular localization. Little expression of IL-1α and IL-1β genes was observed in normal retina, but this was highly upregulated after ischemia and subsequent reperfusion, in a time-dependent manner.[20] Time courses of IL-1α and IL-1β mRNA expression were also different, in that induction of IL-1α mRNA happened before that of IL-1β mRNA. For IL-1β, three types of cells were identified as the cellular origin of its mRNA. One cell type consisted of neutrophils recruited into the retina, and the other two cell types were resident retinal cells, namely astrocytes and endothelial cells.[20] The authors speculated that the neutrophils recruited after ischemia are activated and then synthesized IL-1, which promotes secretion of products that damage microvasculature and retinal tissue.[21]

In a rat model of lung injury obtained by intratracheal instillation of bleomycin, which subsequently leads to fibrosis, Sakanashi et al[22] investigated the kinetics of infiltration of the various macrophage subpopulations. Shortly after the intratracheal instillation of bleomycin the number of exudate macrophages in the lungs increased, peaked 3 days later, and decreased thereafter, whereas tissue macrophages increased slowly and peaked 2 weeks after instillation. To elucidate the molecular mechanisms underlying macrophage infiltration, the mRNA expression of MCP-1 by the alveolar and interstitial cells was examined. Northern blot analysis revealed that the expression of MCP-1 mRNA in the lung was most prominent the first day after instillation and declined thereafter, thus preceding the numerical change of the exudate monocytes. Immunochemistry disclosed that the main sources of MCP-1 production were alveolar and interstitial macrophages, as well as polymorphonuclear neutrophils.[22] Whereas in the early phase of bleomycin-induced lung injury alveolar cells showed intense positivity of MCP-1 mRNA, the high expression of MCP-1 mRNA in the alveolar cells on day 1 was partly explained by the production of MCP-1 by PMN, because these cells accounted for 66% of alveolar cells at this stage. Based on these results, the authors speculated that MCP-1 produced by PMN and by alveolar and interstitial macrophages induced the infiltration of blood monocytes in the very early phase, and that the subsequent accumulation of macrophages is enhanced by the MCP-1 production by monocyte-derived exudate macrophages.[22]

Reminiscent of the contrasting in vitro observations on the presumed ability of neutrophils to express MCP-1,[23-25] there also exist discordant findings on this issue in vivo. D'Angio et al[26] reported that in rabbit lung lavage cells exposed to hyperoxia, the levels of both

IL-8 and MCP-1 mRNA were elevated, suggesting that the two chemokines play important roles in the recruitment of proinflammatory cells. A quantitative in situ hybridization showed that both IL-8 and MCP-1 were expressed in AM, whereas only IL-8 was present in recruited PMN. Interestingly, IL-8 mRNA production in lavage PMN was elevated throughout the time that PMN were available for analysis, and although no data on IL-8 protein were produced, the presence of increased levels of IL-8 mRNA in PMN entering the alveolus implies an autocrine role for this cytokine in PMN activation. Further work addressing the presence of MCP-1 protein in vitro and in vivo is needed to understand whether neutrophils are able to express the chemokine.

Finally, during the course of generalized immune complex-mediated peritonitis in normal mice, Zhang et al,[27] observed that there were two peaks of TNFα secretion into the peritoneal exudate. The first peak occurred 5 min after complex deposition and clearly resulted from the release of TNFα from mast cell granules. Mast cells are in fact the only cell type known to store a preformed TNFα in secretory granules, ready to be rapidly released.[28] The second peak occurred between 4 and 8 h after injury and consisted of newly synthesized TNFα. Indirect evidence suggested that neutrophils, recruited as a result of mast cell-derived early TNFα, might have been the major source of the second peak of TNFα. The ability of TNFα in directly eliciting the initial influx of PMN in the early phase of inflammation is, in fact, well established. The observed influx of neutrophils was subsequently demonstrated to be due to an IL-8-like protein,[29] which could be derived by the same mast cells, or produced by other cells (fibroblasts, endothelial cells), in response to the mast cell-derived TNFα.[29]

NEUTROPHIL-DERIVED CYTOKINES DURING IN VIVO INFECTIONS

To better understand the pathogenesis of *E.coli*-dependent arthritis, Matsukawa et al[30] used a model of rabbit arthritis induced by intra-articular injection of LPS. These authors observed that IL-1β produced by infiltrating neutrophils was responsible for the ensuing tissue destruction.[30] However, because IL-1β in itself has no clear chemotactic capacity for neutrophils, and because a variety of studies done in animal models in vivo had suggested the involvement of IL-8 in recruiting PMN, they decided to elucidate the role of IL-8 in the early stage of arthritis. Using a different experimental system, the same group injected homologous IL-8 in rabbit knee joints and investigated the subsequent inflammatory response.[31] In these experiments, IL-8 induced a massive accumulation of neutrophils (but no appreciable numbers of lymphocytes) and provoked the release of neutrophil elastase, which led to cartilage destruction. In addition, injection of IL-8 induced bioactive and immunoreactive IL-1β and IL-1ra in the joint cavity, but not TNFα.

As determined by immunohistochemistry, IL-1β- and IL-1ra-positive cells were infiltrating leukocytes.[31] Production kinetics of immunoreactive IL-1ra in synovial fluid (SF) overlapped that of IL-1β, but the peak concentration of IL-1ra exceeded that of IL-1β by a 40-50 fold molar ratio. Strikingly, in neutrophil-depleted rabbits, IL-8 induced no cartilage destruction and far lesser concentrations of IL-1β and IL-1ra as compared with normal rabbits,[31] proving that infiltrating neutrophils were the main producers of these cytokines and that they were responsible for cartilage destruction. IL-8 induced little macrophage/lymphocyte accumulation in neutrophil-depleted rabbits, suggesting that early neutrophil accumulation may affect the later accumulation of macrophages or lymphocytes through the production of specific chemoattractans, which the authors did not attempt to identify (MIP-1α/β? MCP-1? IP-10?). Matsukawa et al[31] thus concluded that IL-8 is a potent neutrophil activator in vivo and may have a crucial role in the biology of inflammation and the pathogenesis of inflammatory processes, including septic arthritis.

In a very interesting study, Jordan et al[32] aimed to determine the endogenous mediators involved in the induction of IL-1ra during bacterial infection, and to characterize the cellular origin of IL-1ra. As a model, they used the oral infection of mice with the enteropathogenic *Yersinia enterocolitica*, because bacteria initially proliferate in the tissue of the terminal ileum, predominantly in the Peyer's patches (PP), which are easily accessible for cellular and molecular analysis. In these sites, the immediate antibacterial host defense is characterized by an infiltration of granulocytes and monocytes. By in situ hybridization, northern blot and immunostaining, Jordan et al[32] found expression of IL-1ra mRNA and synthesis of IL-1ra in PP, as well as in uninfected organs such as spleen, but not in the liver, with higher yields on day 3 through day 6. In contrast, the mRNA for IL-1β in PP was expressed considerably earlier, as sessile macrophages were its primary source. Interestingly, no temporal differences were observed between IL-1α and IL-1ra.[32] Circulating and recruited neutrophils, but not PBMC, were identified to be the primary source of IL-1ra in tissues, whereas approximately 20% of the positive IL-1ra-staining cells were accounted for by inflammatory macrophages. In addition, in situ hybridization of adjacent sections of PP revealed a distinct hybridization pattern for each cytokine, suggesting that IL-1α, IL-1β and IL-1ra were produced independently by different cell types, or alternatively by cells of the same phenotype located within different tissue areas. Strikingly, neutralization with an antiserum of IL-6, a cytokine which was also promptly induced by *Y.enterocolitica* infection, caused a suppression of both IL-1ra mRNA in PP, and synthesis of IL-1ra in circulating neutrophils. In support of these findings in vivo, IL-6 induced IL-1ra expression in cultures of macrophages and PMN in vitro, and anti-IL-6 antiserum blocked these effects of IL-6.[32] In this respect, previous studies in humans demonstrated that IL-6 infused into cancer patients rapidly increased the levels of circulating IL-1ra.[33] Altogether, the observations of Jordan et al[32] uncovered important inter-relationships among IL-1, IL-6 and IL-1ra in *Y.enterocolitica* infections. For example, after the production of IL-1 and IL-6 early after *Yersinia* infection, IL-6 in turn induces IL-1ra, which then may inhibit IL-1 activities through a negative feedback loop, facilitating the resolution of the inflammatory response locally and presumably at remote sites of infection.

To gain more insights in the mechanisms underlying neutrophil influx in the airways, during chronic bacterial infection, Inoue et al[34] not only examined the localization of IL-8 mRNA expression after incubating human and dog bronchi with *P.aeruginosa* supernatant in vitro, but they also studied the effect of *P.aeruginosa* supernatant delivered in the dog trachea in vivo on the expression and localization of IL-8 mRNA in airways. The latter procedures caused IL-8 mRNA expression in epithelial and gland duct cells but also in recruited neutrophils.[34] Interestingly, the molecule responsible for this IL-8 induction was not the purified *P.aeruginosa* lipopolysaccharide, but a small molecular weight (1 kDa) product of *P.aeruginosa*.[34] IL-8 expression in recruited neutrophils might provide a potential mechanism for amplifying the inflammatory response and for a positive feedback of a protective antibacterial response, for example by rendering phagocytosis more effective. The latter observations might be relevant in the context of cystic fibrosis, since *Pseudomonas* commonly infects airways of affected patients.

One of the most interesting study on the production of cytokines by neutrophils in vivo is, in my opinion, the one recently conducted by Romani et al.[1] These researchers investigated the role of neutrophils in the generation of murine T helper responses to *C.albicans*. It is in fact well known that subsets of CD4+ T helper cells can be characterized on the basis of their pattern of cytokine production either in mouse[35] or human systems[36]: T helper 1 (Th1) cells, which predominantly produce IL-2, IFNγ and lymphotoxin, that are effective inducers of delayed type hypersensitivity (DTH); and T helper 2 (Th2), mainly producing IL-4, IL-5 and IL-10, that provide more effective help for B cells. Human Th1-like cells preferentially develop during infections by intracellular bacteria, protozoa, and viruses, whereas Th2-like cells predominate during helmintic infestations and in response to common environmental allergens. Strongly polarized human Th1-type and Th2-type responses not only play different roles in protection, but they can also promote different immunopathological reactions.[36] In addition, an altered profile of lymphokine production may account for immune dysfunctions in some primary or acquired immunodeficiency syndromes.[36]

C.albicans is a commensal microorganism that, especially in immunonocompromised hosts, may represent an important cause of morbidity and mortality. Studies in mice have clearly established that multiple mechanisms may control the outcome of experimental infections. For example, in immunized mice such outcome is greatly conditioned by the type of predominant T helper cell subset activated by the initial exposure to the yeast: Th1 cell activation leads to resistance and onset of durable protection, whereas Th2 cell responses are associated with susceptibility to progressive disease. Previous studies in Romani's lab indicated that numerous factors are involved in the preferential induction of murine Th1 or Th2 cell responses to *Candida*.[37] Cytokines emerged, obviously, as key regulators in the development of CD4+ subsets from precursor Th cells,[38,39] and for the Th1-responses several evidence indicated production of IL-12 as fundamental.[40] Importantly, depletion of granulocytes in resistant mice led to the onset of Th2 rather than Th1 responses, indicating that the latter cells may participate in *Candida* driven Th1-development.[37] Interestingly, neutropenia constitutes one of the major factor responsible for fungal dissemination to visceral organs. As described in chapter 4, Romani et al first demonstrated that following stimulation in vitro with either IFNγ plus LPS or different strains of *C.albicans*, granulocytes from naive mice were able to release IL-12 and IL-10 (personal communication). Subsequently, the ability of neutrophils to release IL-12 and IL-10 in vivo, during the course of *C.albicans* infection, was investigated by injecting intravenously the live vaccine strain PCA-2, which causes healing infection, or the highly virulent CA-6 strain which causes nonhealing infection. Under those conditions, IL-12 specific transcripts were consistently expressed by granulocytes in vivo from healer mice, while IL-10 was detected only in mice with progressive disease.[1] Neutrophil depletion prevented the development of protective Th1 responses in healer mice, but increased resistance later in infection of susceptible hosts, the latter finding being related to a decreased IL-10 production. Protective Th1 immunity was efficiently restored by IL-12 administration in neutropenic mice, consistent with a role for neutrophil-derived IL-12 in Th1 development. Another very important observation in the study of Romani et al[1] was that the balance between IL-10 and IL-12 productions by neutrophils was modified by exogenous IL-12, in that PMN-release of IL-10 increased after IL-12 treatment in both uninfected and infected mice. Although this IL-12-induced production of IL-10 by neutrophils might have been the result of indirect mediators stimulating PMN, this mechanism could act as a regulatory response to challenge with IL-12. In addition, such an effect by IL-12 might account for an observation previously made by the same group,[40] of a paradoxical effect of IL-12 in the resistant host. They in fact reported that administration of IL-12 not only fails to promote (enhance) protective anticandidal immunity in nongranulocytopenic mice, but actually promoted Th2 development in a healing infection with detectable levels of circulating IL-10/IL-4.[40] The increased production of IL-10 by neutrophils after IL-12 treatment might be the explanation for or could contribute to the failure of IL-12 to exert protective effects in mice with candidiasis.[1] In conclusion, the results of Romani et al[1] are very important because they demonstrate that PMN, through the release of IL-12 and IL-10 might significantly contribute to the patterns of susceptibility and resistance in mice with candidiasis. More strikingly, the work of Romani et al[39] reports for the first time that neutrophils, via their ability to release cytokines, may play an active role in determining Th selection. The latter concept was previously hypothesized from the work of my lab.[41]

FURTHER EXPERIMENTAL SITUATIONS OF IN VIVO NEUTROPHIL-DERIVED CYTOKINES

Other studies have suggested that the potential production of cytokines by PMN might also significantly affect other processes such as immune and antitumor responses, and bone development.

In a series of papers published by Sendo and coworkers[42-45] for example, cell-mediated immune responses and antibody production were analyzed in rats depleted of PMN using a specific anti-neutrophil monoclonal antibodies designated RP-3. Those experiments demonstrated that both the priming and the elicitation phases of delayed type hypersensitivity (DTH) to sheep red blood cells (SRBC),[42] and the accompanying mononuclear cell recruitment,[43] were partially inhibited in PMN-depleted rats, suggesting that neutrophils enhance DTH to SRBC. Of great relevance, the same group previously demonstrated that IL-8-induced CD4+ T lymphocyte recruitment into subcutaneous tissues of rats were inhibited by the RP-3 treament.[44] Furthermore, by assessing the direct or indirect splenic plaque-forming cell (PFC) response to SRBC in rats depleted of PMN 6-12 h before immunization, they detected an increased number of anti-SRBC antibody-producing cells.[45] This phenomenon was observed only when the antigen was administered intraperitoneally and not with intravenous immunization,[45] and suggested that neutrophils could suppress antibody production in certain situations. Of great interest, using a similar experimental animal model, the same group also demonstrated that transplantation immunity against cancer and generation of CD8+ effector T cells to tumor-associated antigens were abrogated by selective depletion of neutrophils.[46,47] Although the precise mechanisms underlying all these phenomena were not molecularly elucidated by Sendo's group, it can be envisaged that they could be related to the lack of PMN-derived cytokines, which would thus affect leukocyte recruitment (IL-8, MIP-1α, MIP-1β, MIP-2), antigen presentation (IL-1), lymphocyte proliferation and activation (IL-1, IL-6, TGFβ) and macrophage activation (TNFα). Moreover, since PMN have been shown to express class II molecules if treated with G-CSF plus either GM-CSF, IL-3 or IFNγ,[48-50] it is possible that they could also function as antigen-presenting cells.

Collectively, these findings make it tempting to speculate that PMN, in specific situations, can influence the balance between humoral and cell-mediated immunity in the early stages of the immune response.

The potential ability of PMN to mediate antitumor activity in vivo has been clearly evidentiated by Stopacciaro et al.[51] They took advantage of the murine colon adenocarcinoma C-26 cell line engineered to release G-CSF (C-26/G-CSF), to study the mechanisms responsible for inhibition of tumor take in syngeneic animals, and of regression of an established tumor in sublethally irradiated mice injected with these cells. Using C-26/G-CSF they identified the cell types that infiltrate the tumor and the cytokines expressed in situ. It was found that inhibition of tumor take and regression of an established tumor in sublethally irradiated mice occurred through different mechanisms. In the former case, PMN were the main cells responsible for inhibiting the take of C26/G-CSF. In the latter one, PMN, macrophages and T cells, including CD8+ T cells which are required for IFNγ-mediated tumor regression, determined the rejecton of a C26/G-CSF nodule initially grown in sublethally irradiated mice. Both depletion of CD8+T cells, or neutralization of IFNγ produced by CD8+T cells resulted in a reduction of PMN number and TNFα expression, and therefore in tumor progression.[51] Notably, as evidenced by immunohistochemistry and in situ hybridization, either newly recruited granulocytes surrounding the injected neoplastic cells, or, in sublethally irradiated mice, the PMN infiltrating the C-26/G-CSF tumor during its initial growing phase, expressed transcripts for IL-1α, IL-1β and TNFα.[51] More recently,[52] the analysis of the phenotypic changes resulting from the cytokines' activities during the rejection of C26/G-CSF, and the inhibition of such changes by anti-cytokine antibodies, indicated that TNFα was instrumental in tumor regression. C-26/G-CSF regressing tumors were characterized by an hemorrhagic necrosis dependent on the infiltrating leukocytes and the cytotoxic cytokines they produced.[52] Complete tumor regression was the result of tumor cell hypoxia following the damage of the tumor microvasculature that was the target of both cytotoxic cytokines (TNFα) and PMN.[52] Locally produced IL-1 and TNFα induced VCAM-1 and E-selectin on tumor vessels, and thus indirectly attracted T lymphocytes.[52] Treatment with monoclonal antibodies to IFNγ or TNFα blocked tumor regression by inhibiting VCAM-1 and E-selectin expression on tumor-associated endothelial cells, and this resulted in a reduced number of infiltrating leukocytes. Thus, whereas tumor inhibition was mediated mainly by PMN, tumor regression occurred because of the cooperation of PMN and T cells, as well as of a combination of cytokines, for which T cell-derived IFNγ and PMN-derived TNFα were necessary.

Lastly, the time course of appearance of the TGFβ and its localization in developing endochondral bone was examined by Carrington and colleagues.[53] These authors used the demineralized matrix-induced bone forming system in rats. For the first time, TGFβ was detected in developing endochondral bone in vivo. Intracellular immunohistochemical localization of TGFβ revealed that the cell types in which TGFβ could be detected varied with the time after implantation of the demineralized matrix: first were inflammatory cells, and then cells in late hypertrophying and calcifying cartilage, the osteoblasts and, interestingly, also bone marrow granulocytes.[53] Therefore, production of TGFβ by granulocytes may contribute to the regulation of ossification during endochondral bone development.

References

1. Romani L, Mencacci A, Cenci E et al. IL-12 as replacement therapy in neutropenic mice with fungal infection. 1996, submitted.
2. Ulich TR, Guo K, Yin S et al. Endotoxin-induced cytokine gene expression in vivo. IV. Expression of interleukin 1-alpha/β and interleukin 1 receptor antagonist mRNA during endotoxemia and during endotoxin-initiated local acute inflammation. Am J Pathol 1992; 141:61-68.
3. Xing Z, Kirpalani H, Torry D et al. Polymorphonuclear leukocytes as a significant source of tumor necrosis factor alpha in endotoxin-challenged lung tissue. AM J Pathol 1993; 143:1009-1015.
4. Xing Z, Jordana M, Kirpalani H et al. Cytokine expression by neutrophils and macrophages in vivo: endotoxin induces TNFα, MIP-2, IL-1β and IL-6 but not RANTES or TGFβ1 mRNA expression in acute lung inflammation. Am J Respir Cell Mol Biol 1994; 10:148-153: erratum Am J Respir Cell Mol Biol 1994; 10:following 346.
5. Huang S, Paulauskis JD, Godleski JJ et al. Expression of macrophage inflammatory protein-2 and KC mRNA in pulmonary inflammation. Am J Pathol 1992; 41:981-988.
6. Williams JH, Patel K, Hatakeyama D et al. Activated pulmonary vascular neutrophils as early mediators of endotoxin-induced lung inflammation. Am J Respir Cell Mol Biol 1993; 8:134-144.
7. Cirelli RA, Carey LA, Fisher JK et al. Endotoxin infusion in anesthetized sheep is associated with intrapulmonary sequestration of leukocytes that immunohistochemically express tumor necrosis factor-α. J Leuk Biol 1995; 57:820-826.
8. Terebuth PD, Otterness IG, Strieter RM et al. Biologic and immunohistochemical analysis of interleukin-6 expression in vivo. Constitutive and induced expression in murine polymorphonuclear and mononuclear phagocytes. Am J Pathol 1992; 140:649-657.
9. Melani C, Mattia GF, Silvani A et al. Interleukin-6 expression in human neutrophil and eosinophil peripheral blood granulocytes. Blood 1993; 81:2744-2749.
10. Chensue SW, Terebuth PD, Remick DG et al. In vivo biologic and immunohistochemical analysis of interleukin-1 alpha, beta and tumor necrosis factor, during experimental endotoxemia: kinetics, kupffer cell expression and glucocorticoid effects. Am J Pathol 1991; 138:395-402.

11. Nill MR, Oberyszyn TM, Ross MS et al. Temporal sequence of pulmonary cytokine gene expression in response to endotoxin in C3H/Hen endotoxin-sensitive and C3H/HeJ endotoxin resistant mice. J Leuk Biol 1995; 58:563-574.
12. Kolls JK, Xie J, Lei D et al. Differential effects of in vivo ethanol on LPS-induced TNFα and nitric oxide production in the lung. Am J Physiol 1995; 268:L991-998.
13. Mori S, Goto K, Goto F et al. Dynamic changes in mRNA expression of neutrophils during the course of acute inflammation in rabbits. Int Immunol 1994; 6:149-156.
14. Goto F, Goto K, Mori S et al. Biosynthesis of interleukin-1β at inflammatory sites in rabbit: kinetics and producing cells. Br J Exp Path 1989; 70:597-606.
15. Mori S, Goto F, Goto K et al. Cloning and sequence analysis of a cDNA for lymphocyte proliferation potentiating factor of rabbit polymorphonuclear leukocytes: identification as rabbit interleukin-1β. Biochem Biophys Res Commun 1988; 150:1237-1243
16. Ivey CL, Williams FM, Collins PD et al. Neutrophil chemoattractants generated in two phases during reperfusion of ischemic myocardium in the rabbit. Evidence for a role of C5a and interleukin-8. J Clin Invest 1995; 95:2720-2728.
17. Engler RL, Peterson MA, Dobbs A et al. Accumulation of polymorphonuclear leukocytes during 3 h experimental myocardial ischemia. Am J Physiol 1986; 251:H93-H100.
18. Williams FM, Kus M, Tanda K et al. Effect of duration of ischaemia on reduction of myocardial infarct sixe by inhibition of neutrophil accumulation using an anti CD18-monoclonal antibody. Br J Pharmacol 1994; 111:1123-1128.
19. Mullane K, Hatala MA, Kraemer R et al. Myocardial salvage induced by REV-5901: an inhibitor and antagonist of the leukotrienes. J Cardiovasc Pharmacol 1987; 10:398-406.
20. Hangai M, Yoshimura N, Yoshida M et al. Interleukin-1 gene expression in transient retinal ischemia in the rat. Invest Ophthalmol Vis Sci 1995; 36:571-578.
21. Nosé PS. Cytokines and reperfusion injury. J Card Surg 1993; 8:305-308.
22. Sakanashi Y, Takeya M, Yoshimura T et al. Kinetics of macrophage subpopulations and expression of monocyte chemoattractant protein-1 (MCP-1) in bleomycin-induced lung injury of rats studied by a novel monoclonal antibody against rat MCP-1. J Leuk Biol 1994; 56:741-750.
23. Strieter RM, Kasahara K, Allen R et al. Human neutrophils exhibit disparate chemotactic gene expression. Biochem Biophys Res Commun 1990; 173:725-730.
24. Van Damme J, Proost P, Put W et al. Induction of monocyte chemotactic proteins MCP-1 and MCP-2 in human fibroblasts and leukocytes by cytokines and cytokine inducers. Chemical synthesis of MCP-2 and development of a specific RIA. J Immunol 1994; 152:5495-5502.

25. Burn TC, Petrovick MS, Hohaus S et al. Monocyte chemoattractant protein-1 gene is expressed in activated neutrophils and retinoic acid-induced human myeloid cell lines. Blood 1994; 84:2776-2783.
26. D'Angio CT, Sinkin RA, LoMonaco MB et al. Interleukin-8 and monocyte chemoattractant protein-1 mRNAs in oxygen-injured rabbit lung. Am J Physiol 1995; 12:L826-L831.
27. Zhang Y, Ramos BF, Jakschick BA. Neutrophil recruitment by tumor necrosis factor from mast cells in immune complex peritonits. Science 1992; 258:1957-1959.
28. Gordon JR, Galli SJ. Mast cell as a source of both preformed and immunologically-inducible TNFα/cachectin. Nature 1990; 346:274-276.
29. Zhang Y, Ramos BF, Jakschick B et al. Interleukin-8 and mast cell generated tumor necrosis factor-α in neutrophil recruitment. Inflammation 1995; 19:119-131.
30. Matsukawa A, Ohkawara S, Maeda T et al. Production of IL-1 and IL-1 receptor antagonist and the pathological significance in LPS-induced arthritis in rabbits. Clin Exp Immunol 1993; 93:206-211.
31. Matsukawa A, Yoshimura T, Maeda T et al. Neutrophil accumulation and activation by homologous IL-8 in rabbits. IL-8 induces destruction of cartilage and production of IL-1 and IL-1 receptor antagonist in vivo. J Immunol 1995; 154:5418-5425.
32. Jordan M, Otterness IG, Ng R et al. Neutralization of endogenous IL-6 suppresses induction of IL-1 receptor antagonist. J Immunol 1995; 154:4081-4090.
33. Tilg H, Trehu E, Atkins MB et al. IL-6 as anti-inflammatory cytokine: induction of circulating IL-1 receptor antagonist and soluble tumor necrosis factor receptor p55. Blood 1994; 83:113-118.
34. Inoue H, Massion P, Ueki IF et al. *Pseudomonas* stimulates Interleukin-8 mRNA expression selectively in airway epithelium, in gland ducts, and in recruited neutrophils. Am J Respir Cell Mol Biol 1994; 11:651-663.
35. Mosmann TR, Coffman RL. Th1 and Th2 cells: different patterns of lymphokine secretion lead to different functional properties. Ann Rev Immunol 1989; 7:145-173.
36. Romagnani S. Lymphokine production by human T cells in disease state. Ann Rev Immunol 1994; 12: 227-257.
37. Romani L, Puccetti P, Bistoni F. Biological role of helper T cell subsets in candidiasis. Chem Immunol 1995, in press.
38. Puccetti P, Romani L, and Bistoni F. A Th1-Th2-like switch in candidiasis: new perspective for therapy. Trends Microbiol. 1995; 3:237-240.
39. Romani L, Mencacci A, Tonnetti L et al. Interleukin-12 but not interferon-γ production correlates with induction of T helper type-1 phenotype in murine candidiasis. Eur J Immunol 1994; 22:909-913.
40. Romani L, Bistoni F, Mencacci A et al. Interleukin-12 in *Candida* albicans infections. Res Immunol 1995; in press.
41. Cassatella MA, Meda L, Gasperini S et al. Interleukin-12 production by human polymorphonuclear leukocytes. Eur J Immunol 1995; 25:1-5.

42. Kudo C, Yamashita T, Araki A et al. Modulation of in vivo immune response by selective depletion of neutrophils using a monoclonal antibody, RP-3. I. Inhibition by RP-3 treatment of the priming and effector phases of delayed type hypersensitivity to sheep red blood cells in rats. J Immunol 1993; 150:3728-3738.
43. Kudo C, Yamashita T, Terashita M et al. Modulation of in vivo immune response by selective depletion of neutrophils using a monoclonal antibody, RP-3. II. Inhibition by RP-3 treatment of mononuclear leukocyte recruitment in delayed type hypersensitivity to sheep red blood cells in rats. J Immunol 1993; 150:3739-3746.
44. Kudo C, Araki A, Yamashita T et al. Inhibition of IL-8 induced W3/25+ (CD4+) T lymphocyte recruitment into subcutaneous tissue of rats by selective depletion of in vivo neutrophils with a monoclonal antibody. J Immunol 1991; 147:2196-2201.
45. Tamura M, Sekiya S, Terashita M et al. Modulation of in vivo immune response by selective depletion of neutrophils using a monoclonal antibody, RP-3. III. Enhancement by RP-3 treatment of the anti-sheep red blood cell plaque-forming cell response in rats. J Immunol 1994; 153:1301-1308.
46. Midorikawa Y, Yamashita T, Sendo F. Modulation of the immune response to transplanted tumor in rat by selective depletion of neutrophils in vivo using a monoclonal antibody: abrogation of specific transplantation resistance to chemical carcinogen-induced syngeneic tumors by selective depletion of neutrophils in vivo. Cancer Res 1990; 50:6243-6247.
47. Tanaka E, Sendo F. Abrogation of tumor-inhibitory MRC-OX+ (CD8+) effector cells T cell generation in rats by selective depletion in neutrophils in in vivo using a monoclonal antibody. Int J Cancer 1993; 54:131-136.
48. Gosselin EJ, Wardwell K, Rigby WFC et al. Induction of MHC class II on human polymorphonuclear neutrophils by granulocyte/macrophage colony-stimulating factor, IFNγ and IL-3. J Immunol 1993; 151:1482-1490.
49. Mudzinski SP, Christian TP, Guo TL et al. Expression of HLA-DR on neutrophils from patients treated with granulocyte-macrophage colony stimulating factor for mobilization of stem cells. Blood 1995; 86:2452-2453.
50. Smith WB, Guida L, Sun Q et al. Neutrophils activated by granulocyte-macrophage colony stimulating factor express receptors for interleukin-3 which mediate class II expression. Blood 1995; 86:3938-3944.
51. Stoppacciaro A, Melani C, Parenza M et al. Regression of an established tumor genetically modified to release granulocyte colony-stimulating factor requires granulocyte-T cell cooperation and T cell-produced IFNγ. J Exp Med 1993; 178:151-161.
52. Colombo MP, Lombardi L, Melani C et al. Hypoxic tumor cell death and modulation of endothelial adhesion molecules in the regression of G-CSF transduced tumors. Am J Pathol 1996, in press.
53. Carrington JL, Roberts AB, Flanders KC et al. Accumulation, localization, and compartmentation of Transforming Growth factor β during endochondral bone development. J Cell Biol 1988; 107:1969-1975.

CHAPTER 9

FINAL REMARKS AND FUTURE DIRECTIONS

Although relatively novel, the research on cytokine production by neutrophils has brought forward new and exciting discoveries, which have shown that PMN can no longer be regarded as cells that only release preformed mediators. However, it must be stressed that no evidence has yet been presented that any of the cytokines produced by neutrophils are unique to this cell type. Thus, polymorphonuclear leukocytes (PMN) are likely to represent one among several potential sources of cytokines. It is still premature to assess the actual biological significance of cytokine production by neutrophils, especially because many aspects need to be extended, clarified or approached ex novo. Nevertheless, the fact that neutrophils clearly predominate over other cell types under various in vivo conditions suggests that, in some circumstances, the contribution of PMN-derived cytokines can be of foremost importance.

DISTINCT PATTERNS OF CYTOKINE RELEASE ELICITED BY DIFFERENT NEUTROPHIL STIMULI

One facet of neutrophil biology, which in my opinion needs to be fully elucidated, is the identification of all the stimuli able to induce cytokine synthesis in neutrophils. This might help to understand the pathogenesis of diseases in whose early phases, neutrophils represent (or are presumed to be) the first cell type encountering, and interacting with, the etiologic agent. For example, it is well established that PMN can secrete a number of different chemokines. So, the influx of the different leukocyte populations to inflammatory lesions might be controlled depending upon which chemokine is produced. IL-8 and GROα will predominantly recruit neutrophils, whereas MIP-1α/β or IP-10 will essentially recruit monocytes and lymphocyte subtypes. Thus, depending upon the nature of the primary insult, and its effect towards the

---

Cytokines Produced by Polymorphonuclear Neutrophils: Molecular and Biological Aspects, by Marco A. Cassatella. © 1996 R.G. Landes Company.

production of chemokines by neutrophils, the evolution of a given type of inflammatory reaction may be anticipated. In this regard, I have already summarized (in several tables) the current knowledge of the ability of different agents to induce specific cytokines. Although for many of them the picture is still incomplete, it is clear that the interaction of PMN with a given agonist produces a characteristic response.

Our investigations have focused mostly on the effects of S.cerevisiae opsonized with IgG (Y-IgG), lipopolysaccharide (LPS), and formyl-methionyl-leucyl-phenylalanine (fMLP) as stimuli. As already outlined in several instances throughout this book, a strikingly different pattern of cytokine production and regulation is observed when Y-IgG-phagocytosis is compared to stimulation with endotoxin. These differences can be summarized as follows. First, Y-IgG phagocytosis is a much more efficient stimulus than LPS for the extracellular release of TNFα, IL-8 and GROα,[1,2,3] but does not induce IL-12p40 mRNA and protein,[4] and represents a very poor inducer of IL-1ra release.[5] Second, since the extent of the up-modulatory effects of Y-IgG on TNFα, IL-8 and GROα mRNA accumulation are equivalent (if not lower) to those exerted by LPS, it is likely that Y-IgG-stimulated cytokine production is controlled by translational and/or secretional processes. We have shown that de novo synthesized IL-8 is more efficiently secreted in Y-IgG-stimulated PMN than in LPS- or TNFα-activated cells.[6] Furthermore, while IFNγ increased the percentage of IL-8 being secreted by PMN in response to LPS or TNFα, it did not influence that observed in response to Y-IgG.[6] Third, the Y-IgG- and LPS-elicited cytokine production are differently influenced by IL-10 and IFNγ. On the one hand, IL-10 slightly inhibits Y-IgG-induced TNFα, IL-1β, IL-8 and GROα release,[3,7] without increasing the production of IL-1ra, as observed in the case of LPS.[5] On the other hand, IFNγ moderately increases Y-IgG-stimulated production of IL-1γ without significantly enhancing that of TNFα,[6] but inhibiting IL-8[6] and GROα[3] production to a various extent. By comparison, IFNγ potentiates IL-1β, TNFα, IL-8 and GROα release induced by a 18 h treatment with LPS.[3,6] Fourth, while LPS in combination with IFNγ induces the extracellular production of both IL-12[4] and IP-10 (manuscript in preparation), Y-IgG does not. Fifth, in the case of Y-IgG-stimulated cells, endogenous TNFα and IL-1β do not play determinant roles for IL-8 production,[6,7] in contrast to LPS-stimulated cells. In view of all of the above differences, it is evident that LPS and Y-IgG-phagocytosis utilize distinct biochemical and molecular intracellular pathways to induce cytokine production in PMN. This in turn gives rise to potentially important questions. For instance, why does Y-IgG-phagocytosis potently trigger the release of determined proinflammatory cytokines (TNFα, IL-8 and GROα), whereas LPS induces all types of pro- and anti-inflammatory cytokines (including IL-12p40 and IL-1ra)? Similarly, why are Y-IgG poor inducers of IL-1ra, and are completely unable to induce IL-12 in neutrophils, whereas they can do so in monocytes? And finally, do these observations reflect what really happens in vivo?

Similar considerations arise if the action of fMLP (or of other chemotactic factors) is examined. As already discussed in chapter 4, fMLP triggers only a transient release of IL-8 and low levels of GROα in PMN, but apparently, no IL-1β, TNFα, IL-12, MIP-1α/β or IP-10 production.[3,8,9] It appears therefore that neutrophil chemotactic factors induce neutrophils to produce chemokines which activate and recruit more neutrophils, in a sort of positive feedback loop. Have these in vitro findings an in vivo counterpart, and, if so, what could be its possible relevance? While it is too early to give definitive answers, some results already obtained in vivo are reminiscent of the in vitro findings. For instance, I described in chapter 8 the model of myocardial infarction generated in the rabbit,[10] which uncovered some of the basic mechanisms underlying neutrophil accumulation in this type of inflammatory response. The authors demonstrated that the process of neutrophil accumulation in myocardial tissue after ischemia and reperfusion could be explained by the sequential production of the complement fragment C5a, which recruited a certain number of neutrophils, which in turn generated IL-8, presumably in response to C5a and other stimuli.[10] These observations, therefore, have uncovered a scenario in which possible relationships between C5a and neutrophil-derived IL-8 exist in vivo.

Other interesting and intriguing problems arise from the findings regarding the production of IL-12 and IP-10 by PMN. The inducible expression of both cytokines requires IFNγ in combination with another stimulus: LPS, in the case of IL-12,[4] and LPS or TNFα, in the case of IP-10 (manuscript in preparation). Apart from the necessity to teleologically explain the reasons why at least two stimuli are required to induce IL-12 or IP-10 (a task already clarified at the molecular level for IL-12[4]), the relevance of these findings should be interpreted in consideration of the fact that IFNγ is one of the two stimuli. Since IFNγ is mainly produced by Th1 type lymphocytes,[11] an eventual production of IL-12 or IP-10 by PMN would take place, in theory, after the initiation of the immune response, and only by those neutrophils which have survived long enough at the inflammatory focus, or which are actively recruited in the context of chronic inflammatory or immune processes. In this scenario, the role of the release of IL-12 or IP-10 by PMN would be to enhance or sustain a cell-mediated immune response, as opposed to influencing the development of the immune response toward a cell-mediated one. While this might hold true, it must be recalled that IFNγ is also made by Natural Killer (NK) cells,[11] which constitute one of the effector cell types of natural immunity. Interestingly, in a model of skin lesion induced by suction (a human experimental inflammatory focus), IFNγ appears one of the first molecules to accumulate, preceding other cytokines such as IL-6, IL-8, IL-1β, TNFα and GM-CSF.[12] Although skin-associated T cells have been suggested to be a likely source of IFNγ,[12] these results demonstrate that under some conditions, IFNγ can be released in the initial phase of the inflammatory response. In such cases, IFNγ could act not only as a potentiating factor of cytokine production by neutrophils,[6] but also as a direct co-inducer of IL-12 or IP-10.

DIFFERENTIAL ABILITIES OF NEUTROPHILS AND MONONUCLEAR CELLS TO PRODUCE CYTOKINES

The extent to which neutrophils (as opposed to other cell types) are an important source of a particular cytokine may vary greatly, and will depend not only on the characteristics of the response under investigation, but also on whether neutrophils constitute the predominant cell type. Studies which compared the respective ability to produce cytokines of PMN and either purified monocytes or peripheral blood mononuclear cells (PBMC) stimulated under similar conditions have revealed that, as expected, the latter cellular types make substantially greater quantities of cytokines than PMN, if evaluated on a single-cell basis.[13] Depending upon the molecules investigated, monocytes/PBMC produce cytokines in amounts that exceed those produced by PMN by a factor of twenty to hundred folds (for a direct comparison of the amounts of specific cytokines released by PMN and either purified monocytes or PBMC, please refer to the table already published in ref. 13). While it is evident that monocytes/PBMC make substantially greater quantities of cytokines than PMN on an individual cell basis, it must be emphasized that, from a physiopathological point of view, granulocytes constitute the majority of infiltrating cells in inflamed tissues, and may thus represent an important source of cytokines under such conditions.

As outlined many times throughout this book, I believe it is important to thoroughly characterize the differential induction of cytokines by neutrophils and mononuclear cells. These differences, which will be briefly summarized below, provide a number of clues which could eventually allow a better understanding of the respective contribution of the above cell types to various inflammatory/immunologically processes. For instance, Figure 9.1 shows a representative Northern blot experiment in which PMN and PBMC isolated from the same donor were pretreated with IL-10 and then stimulated for 5 h with LPS. It is evident, first of all, that neutrophils do not express IL-6 mRNA in any conditions. Interestingly, while the levels of IL-8 and IL-1ra mRNA are relatively more abundant in LPS-treated neutrophils than in PBMC, the opposite is observed for TNFα, IL-1β and IL-12p40. Furthermore,

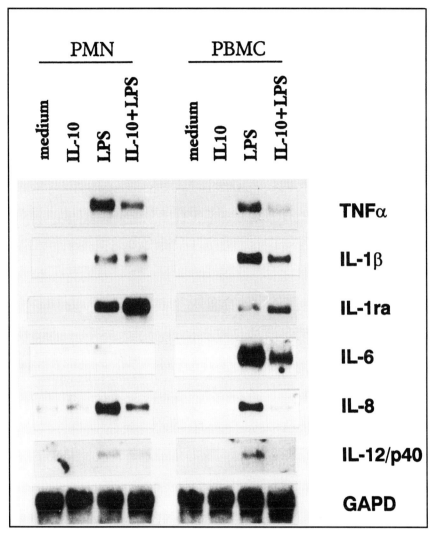

Fig. 9.1. Effect of IL-10 on cytokine mRNA accumulation in neutrophils and peripheral blood mononuclear cells (PBMC) stimulated for 5 h with LPS.

while IL-10 has little effect on LPS-induced IL-1β mRNA levels in PMN, it effectively inhibits these transcripts in monocytes.

Interferon-α (IFNα)

G-CSF specifically stimulates neutrophils, but not monocytes, to express and release IFNα.[14] IFNα1, IFNα2 and IFNα4 transcripts constitute the major RNA species present in PBMC and PMN.[15] In monocytes, however, IFNα5 transcripts are also present, and constitute the major IFNα species.[15] While no antiviral activity is detected in culture supernatants of uninduced cell populations, both PMN and monocytes, following induction with Sendai virus, express greater amounts of the transcripts encoding all the IFNα subtypes, and release equivalent amounts of acid-stable IFNα protein.[15] These results suggest that the differential expression of particular IFNα subtypes in these cells might help in understanding the biological role of IFNα subtypes.

Transforming Growth Factor-β (TGFβ)

Neutrophils constitutively secrete an active form of TGFβ, whereas monocytes do not.[16] Secretion of TGFβ by monocytes is however stimulated by LPS, at levels comparable to those made by an equivalent number of resting neutrophils.[16] Interestingly, PMN constitutively express higher levels of TGFβ1 mRNA than monocytes, regardless of whether the latter are activated by LPS.[16]

Tumor Necrosis Factor-α (TNFα)

The LPS-dependent TNFα production by human neutrophils was increased by exposure to sodium nitroprussiate (SNP)(a drug which is employed as an exogenous source of nitric oxide), both in the presence or absence of N-acetylcysteine (NAC)(which increases the bioavailability of nitric oxide).[17] SNP plus NAC also potentiated the up-regulatory effect of IFNγ on LPS-induced TNFα production.[17] In contrast, addition of SNP plus NAC did not produce any effect on the release of TNFα by cultured monocyte-derived macrophages treated with a wide range of LPS doses (0.1-1000 ng/ml), in the presence or absence of IFNγ.[17]

Interleukin-1

In a study in which the cellular source of IL-1β in LPS-stimulated whole blood was examined by immunohistochemistry, monocytes displayed a marked increase in staining intensity by 4 h, whereas PMN expressed low-level positivity with no increase over time; lymphocytes remained negative throughout however.[18] The authors suggested that monocytes are the major producers of IL-1β in a model that closely mimics the in vivo state.[18] Conversely, in a model of casein-induced peritoneal inflammation in rabbits, both PMN and macrophages were found to produce IL-1β, yet neutrophils were the paramount producers of IL-1β;[19] at no time point were macrophages the major IL-1β producer in this type of inflammation.[19] These data[18,19] illustrate the important notion that depending upon the type of inflammation, the cell types contributing to the production of IL-1β (or other cytokines) can greatly vary.

Lord et al[20] observed that while the LPS-elicited increases in IL-1 mRNA in both PMN and PBMC were similar, PMN were much less efficient than PBMC in translating IL-1 mRNA. However, IL-1α and

IL-1β mRNAs purified from PMN and PBMC were translated with equal efficiency in rabbit reticulocyte lysates[20,] suggesting a translational regulation of IL-1 mRNA in mature PMN.

INTERLEUKIN-1 RECEPTOR ANTAGONIST (IL-1RA)

Ulich et al[21] showed that mononuclear cells and PMN express equivalent amounts of IL-1ra mRNA, after incubation in vitro with LPS for 4 h. This was also confirmed in other laboratories[22] (see also Fig. 9.1). In agreement with previous observations,[23,24] Ulich et al[21] also reported that in monocytes, IgG-coated culture dishes proved to be a stronger stimulus than LPS for IL-1ra mRNA upregulation. In contrast, Malyak et al[25] found that culture of PMN on plates coated with adherent human IgG failed to stimulate IL-1ra synthesis in PMN. Similarly, we observed that Y-IgG were three to five times less efficient than LPS in inducing IL-1ra mRNA and protein release in human neutrophils[5] (Figs. 3.2 and 4.1). Apart from the evidence that in all these studies monocyte contamination was not responsible for the observed IL-1ra production attributed to neutrophils, these findings indicate that stimulation through Fcγ-receptors has fundamentally different consequences.

IL-10 alone induces a small accumulation of IL-1ra mRNA and protein in PBMC, but not in neutrophils,[5,26,27] whereas IL-13 is able to induce the production of IL-1ra in both cell types.[28] In the latter experimental conditions, secreted IL-1ra accounts for 72% and 32% of total synthesized IL-1ra by monocytes and PMN, respectively. Lately, Muzio et al[29] cloned a new isoform of intracellular IL-1ra, referred to as type II icIL-1ra (icIL-1raII). When the expression of icIL-1raII was studied in monocytes by reverse-transcriptase polymerase chain reaction (RT-PCR), its mRNA was induced by LPS, but not by IL-4, IL-10, IL-13 and aggregated IgG, which were instead equally effective in inducing icIL-1raI expression. In neutrophils, icIL-1raII was constitutively expressed, and was not significantly modified by any of the above stimuli.[29] Thus, the pattern of induction of icIL-1raII is different in monocytes and PMN.

INTERLEUKIN-6 (IL-6)

If human neutrophils and monocytes are purified from the same donors and stimulated in vitro with either LPS or Y-IgG, the expression of the IL-6 gene and its production is observed only in monocytes[1] (see also Figs. 3.4 and 9.1), provided that the PMN preparations are purified enough. With LPS as a stimulus, similar results were reported using in situ hybridization,[30] Northern blot analysis[31] or using RT-PCR.[32]

INTERLEUKIN-8 (IL-8)

Fujishima et al[33] analyzed equal amounts of total mRNA derived from monocytes and neutrophils, and found that constitutive IL-8 mRNA levels were higher in neutrophils than in monocytes, and that after

LPS stimulation both types of cells showed comparable levels of IL-8 mRNA. This was identical to what we had previously published,[2] and is also illustrated in Figure 9.1. With respect to the production of antigenic IL-8 however, it was calculated that a single monocyte secretes 70-fold more IL-8 than does a single neutrophil after 4 h of incubation with LPS.[33] In another work, analysis of total IL-8 production by PBMC following a 24 h stimulation with C5a resulted in approximately 163 ng/ml of IL-8, most of it being released.[34] On the contrary, the total quantity of IL-8 produced by PMN (intracellular plus extracellular) in response to C5a was approximately eight-fold greater than extracellular alone (6.4 ng/ml vs. 0.79 ng/ml), suggesting that PBMC secrete IL-8 more efficiently than PMN.

While peripheral blood monocytes were potently induced by GM-CSF and IL-3 to produce IL-8 mRNA and protein, neutrophils responded only to GM-CSF.[35] In keeping with the observations with C5a,[34] time course studies indicated that IL-8 induced by GM-CSF in PMN was initially cell-associated and later partially secreted. In contrast, in monocytes, GM-CSF-induced IL-8 was entirely secreted from the beginning.[35] Furthermore, Konig et al,[36] evaluated IL-8 release by PMN and PBMC in the context of *P.aeruginosa* infection, and reported that in contrast to PBMC, intracellular IL-8 in PMN exceeded the IL-8 release in unstimulated as well as in stimulated cells, by up to 10-fold. Altogether, these data may reflect several possibilities: a more rapid packaging and secretion of IL-8 in monocytes; an incomplete release or a significant receptor-mediated internalization of IL-8 occurring in PMN; the existence of another molecule released by PBMC, which is antigenically similar to IL-8, and thus is aspecifically recognized by anti-IL-8 antibodies; and degradation of IL-8 by neutrophil-derived proteases.

In contrast with previous work showing that prostaglandin E_2 (PGE_2) was a potent inhibitor of monocyte-derived IL-8,[37] PGE_2 failed to suppress PMN-derived IL-8.[38] Interestingly, PGE_2 also failed to inhibit IL-8 and TNFα release from alveolar-macrophages.[40] It might be possible that, for unknown reasons, both PMN and alveolar-macrophages have lost the ability to regulate IL-8 expression in response to PGE_2.

Figure 9.2 shows the effects of IL-10 and IFNγ on the production of IL-8 by neutrophils and monocytes isolated from the same donor and stimulated for 20 h with LPS, Y-IgG and TNFα. Again, significant differences can be observed. For example, IL-10 inhibits the production of IL-8 determined by LPS more strongly in neutrophils than in monocytes, but the opposite occurs if the stimulus is TNFα, in keeping with the lack of effect of IL-10 towards TNFα-elicited PMN responses.[3,7,40] Conversely, IFNγ potentiates the release of IL-8 induced by LPS and TNFα in neutrophils, but inhibits that induced in monocytes by the same stimuli. Furthermore, IFNγ inhibits Y-IgG induced IL-8 release from PMN, but not from monocytes (Fig. 9.2).

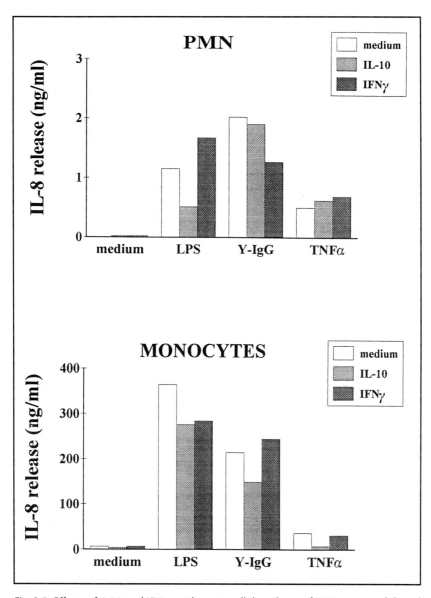

Fig. 9.2. Effects of IL-10 and IFNγ on the extracellular release of IL-8 in neutrophils and monocytes stimulated under different conditions. PMN (5 × 10^6/ml) and monocytes (5 × 10^5/ml) isolated from the same donor were treated with IL-10 and IFNγ and then cultured for 18 h with LPS, Y-IgG, or TNFα (5 ng/ml), before measuring IL-8 protein in their cell-free supernatants.

GROWTH RELATED GENE PRODUCT-α (GROα)

Haskill et al,[41] utilizing PCR primer pairs specific for each of the three GRO cDNAs, observed that neutrophils adherent to fibronectin produced only GROα, lymphocytes produced low but significant levels of GROβ only, whereas monocytes expressed all three GRO messages. We also found that GROα production is differently regulated in neutrophils and monocytes of the same donor.[3] In fact, Y-IgG-phagocytosis was more potent than LPS and TNFα in inducing GROα extracellular release from neutrophils, but in monocytes, the most potent stimulus was LPS.[3] Furthermore, although IL-10 down-regulated the LPS-induced GROα production in both neutrophils and monocytes, in the former cells this effect occurred in the absence of a diminution of GROα mRNA levels, whereas in monocytes IL-10 down-regulated LPS-induced GROα messages.[3]

INTERLEUKIN-12 (IL-12)

As described in chapter 4, we have shown that neutrophils are able to produce biologically active IL-12p70 only in response to LPS plus IFNγ, while in monocytes LPS was effective by itself.[4] Figure 4.3 reports another difference between PMN and monocytes in relation to the ability of Y-IgG to induce IL-12 production. Remarkably, neutrophils isolated from the peritoneal cavity of naive mice and cultured in vitro in the presence of LPS plus IFNγ release IL-12p70 (L.Romani, personal communication). Furthermore, assessment of cytokine gene expression by RT-PCR revealed that neutrophils stimulated for 2 h with IFNγ plus LPS show an accumulation of the messages for IL-12p40, IL-6, IL-10, and TNFα, while in peritoneal macrophages, the mRNA for IL-6 and TNFα were detected, but not those encoding IL-10 and IL-12p40 (L. Romani, personal communication).

INTERFERON INDUCIBLE PROTEIN-10 (IP-10)

Similar to the situation observed with IL-12, we found that neutrophils release IP-10 only in response to IFNγ plus LPS, or IFNγ plus TNFα. In contrast, monocytes produce IP-10 if stimulated by IFNγ only (manuscript in preparation).

STEM CELL FACTOR (SCF)

Ramenghi et al,[42] by using RT-PCR, observed that, in human peripheral blood, SCF expression could be ascribable to PMN, whereas no SCF expression was detected in isolated lymphocytes or monocytes.

INTRACELLULAR CONTROL OF CYTOKINE PRODUCTION

A very important aspect which, to date, has been mostly ignored by investigators working in the field of cytokine production by neutrophils is the potential involvement of transcription factors in the regulation of cytokine gene transcription. This partially reflects the fact that very little is known about the transcriptional events which control cytokine and chemokine gene expression in PMN. However, the inhibitory effects of actinomycin D towards the inducible accumulation of cytokine mRNA in PMN, and a limited number of direct evidence (see chapter 6), suggests that transcriptional events might play a central role in such processes. Many of the regulatory elements located in the promoter regions of most cytokines, as well as the families of transcription factors which bind to them and control their transcription, have been already well identified and characterized in various cell types. However, there can exist substantial differences in the exact pattern of transcription factor binding, depending upon the cell type. Thus, performing such studies in PMN could represent a step forward in our understanding of the cell-specific regulation of cytokine gene expression. Among other things, these studies could lead to the identification of novel transcription factors, and eventually neutrophil-specific factors. Moreover, they could potentially elucidate the molecular basis of the many qualitative and quantitative differences observed between neutrophils and monocytes in their ability to produce determined cytokines, such as IL-6, IL-12, IP-10, and so forth. Finally, the expression of neutrophil transcription factors could be the result of a regulated myeloid differentiation program, which in certain hematopoietic diseases or malignancies may be altered. Consequently, we are currently investigating the activation of selected families of transcription factors by inflammatory agonists in human neutrophils, and, despite the serious technical difficulties initially encountered, our preliminary results are encouraging and suggest that it is well worth conducting such studies.

In a broader context, studies addressing the mechanisms that regulate the intracellular distribution and release of cytokines are very scarce, both in neutrophils and other cell types. The need for such studies is great given that they would advantageously complement the considerable knowledge accumulated on cytokine gene and protein expression.

CONCLUSION

I have already mentioned that the classical role attributed to neutrophils in immunology textbooks (i.e., a cell type whose predominant function is the phagocytosis and killing of bacteria) is still based on the obsolete view that PMN are terminally differentiated, short-lived cells, with minimal, if any, transcriptional or translational activity. Accordingly, the importance of this cell type is usually illustrated using clinical cases (patients with reduced neutrophil numbers, or with rare genetic defects of neutrophils, who are more susceptible to infections than normal individuals) (see for example ref. 43). However, studies conducted in the last six to seven years have forced a re-evaluation of the actual role of neutrophils in host-defense processes. The fundamental contribution of these studies, which I have attempted to describe as comprehensively as possible in this book, has been to clearly demonstrate the ability of neutrophils to synthesize and release various cytokines. This new dimension of neutrophil biology has opened new perspectives as to the potential role of these cells in the context of both inflammmatory and immune responses. In addition, an increasing body of evidence, mainly deriving from in vitro studies,[44] indicates that PMN survival can be greatly extended following exposure to microenvironmental signals, such as LPS, inactivated *Streptococci*, IL-1β, TNFα, IL-6, IFN-γ, G-CSF, and GM-CSF.[45,46] Indeed, these observations therefore raise the possibility that PMN viability in vivo may be considerably greater than what has been heretofore believed. If so, the ability of neutrophils to synthesize immunomodulatory cytokines could prove to be a phenomenon of considerable pathophysiological importance. In any event, it has become clear that PMN should be considered not only as active and central elements of the inflammatory response, but also as cells that, through cytokine secretion, may significantly influence the direction and evolution of inflammatory and immune processes.

In view of the variety of cytokines and chemokines secreted by neutrophils, it can be envisaged that PMN can orchestrate the infiltration of leukocytes into sites of injury, and therefore determine the evolution of the host response. In this scenario, PMN would play a pivotal role in the regulatory interactions between innate resistance (mediated by phagocytic cells and NK cells) and adaptive immunity (mediated by T and B cells). Although there is no published evidence showing that PMN can process or present antigens to T cells, recent studies have indicated that PMN can synthesize and express MHC class II molecules on their surface.[47-49] This raises the possibility that PMN might also initiate a cellular immune response. This putative function of neutrophils would imply a potential role of these cells in many pathological conditions.

Even though our understanding of cytokine production by PMN is far from complete, particularly in humans in vivo, its full appreciation is likely to yield new insights into therapy of many disorders known to be influenced by PMN. In vivo studies will also be essential for critically testing specific hypotheses about the biological significance of neutrophil cytokine production in health and disease.

REFERENCES

1. Bazzoni F, Cassatella MA, Laudanna C et al. Phagocytosis of opsonized yeast induces TNFα mRNA accumulation and protein release by human polymorphonuclear leukocytes. J Leuk Biol 1991; 50:223-228.
2. Bazzoni F, Cassatella MA, Rossi F et al. Phagocytosing neutrophils produce and release high amounts of the neutrophil activating peptide 1/Interleukin 8. J Exp Med 1993; 173:771-774.

3. Gasperini S, Calzetti, F, Russo MP et al. Regulation of GROα production in human granulocytes. J Inflamm 1995; 45:143-151.
4. Cassatella MA, Meda L, Gasperini S et al. Interleukin-12 production by human polymorphonuclear leukocytes. Eur J Immunol 1995; 25:1-5.
5. Cassatella MA, Meda L, Gasperini S et al. Interleukin 10 upregulates IL-1 receptor antagonist production from lipopolysaccharide-stimulated human polymorphonuclear leukocytes by delaying mRNA degradation. J Exp Med 1994; 179:1695-1699.
6. Meda L, Gasperini S, Ceska M et al. Modulation of proinflammatory cytokine release from human polymorphonuclear leukocytes by gamma interferon. Cell Immunol 1994; 57:448-461.
7. Cassatella MA, Meda L, Bonora S et al. Interleukin 10 inhibits the release of proinflammatory cytokines from human polymorphonuclear leukocytes. Evidence for an autocrine role of TNFα and IL-1β in mediating the production of IL-8 triggered by lipopolysaccharide. J Exp Med 1993; 178:2207-2211.
8. Cassatella MA, Bazzoni F, Ceska M et al. Interleukin 8 production by human polymorphonuclear leukocytes. The chemoattractant formyl-Methionyl-Leucyl-Phenylalanine induces the gene expression and release of interleukin 8 through a pertussis toxin sensitive pathway. J Immunol 1992; 148:3216-3220.
9. Contrino J, Krause PJ, Slover N et al. Elevated interleukin 1 expression in human neonatal neutrophils. Pediatr Res 1993; 34:249-252.
10. Ivey CL, Williams FM, Collins PD et al. Neutrophil chemoattractants generated in two phases during reperfusion of ischemic myocardium in the rabbit. Evidence for a role of C5a and interleukin-8. J Clin Invest 1995; 95:2720-2728.
11. Farrar MA, Schreiber RD. The molecular cell biology of Interferon-γ and its receptor. Ann Rev Immunol 1993; 11:571-611.
12. Kuhns D, DeCarlo E, Hawk DM et al. Dynamics of the cellular and humoral components of the inflammatory responses elicited in skin blisters in humans. J Clin Invest 1992; 89:1734-1740.
13. Cassatella MA. The production of cytokines by polymorphonuclear leukocytes. Immunol Today 1995; 16:21-26.
14. Shirafuji N, Matsuda S, Ogura H et al. Granulocyte-Colony-stimulating factor stimulates human mature neutrophilic granulocytes to produce interferon-α. Blood 1990; 75:17-19.
15. Brandt ER, Linnane AW, Devenish RJ. Expression of IFNα genes in subpopulations of peripheral blood cells. Br J Hematol 1994; 86:717-725.
16. Grotendorst GR, Smale G, Pencev D. Production of Transforming growth factor beta by human peripheral blood monocytes and neutrophils. J Cell Physiol 1989; 140:396-402.
17. Van Dervort AL, Yan L, Madara PJ et al. Nitric oxide regulates endotoxin-induced TNFα production by human neutrophils. J Immunol 1994; 152:4102-4109.

18. Hsi ED, Remick DG. Monocytes are the major producers of Interleukin-1β in an ex vivo model of local cytokine production. J Interferon and Cytokine Research 1995; 15:89-94.
19. Goto F, Goto K, Mori S et al. Biosynthesis of interleukin-1β at inflammatory sites in rabbit: kinetics and producing cells. Br J Exp Path 1989; 70:597-606.
20. Lord PCW, Wilmoth LMG, Mizel SB et al. Expression of interleukin-1 alpha and β genes by human blood polymorphonuclear leukocytes. J Clin Invest 1991; 87:1312-1321.
21. Ulich TR, Guo K, Yin S et al. Endotoxin-induced cytokine gene expression in vivo. IV. Expression of interleukin 1-alpha/β and interleukin 1 receptor antagonist mRNA during endotoxemia and during endotoxin-initiated local acute inflammation. Am J Pathol 1992; 141:61-68.
22. Re F, Mengozzi M, Muzio M et al. Expression of interleukin 1 receptor antagonist by human circulating polymorphonuclear cells. Eur J Immunol 1993; 23:570-573.
23. Poutsiaka DD, Clark BD, Vannier E et al. Production of IL-1 receptor antagonist and IL-1β by peripheral blood mononuclear cells is differentially regulated. Blood 1991; 78:1275-1281.
24. Arend PW, Smith MF Jr, Janson RW et al. IL-1 receptor antagonist and IL-1β production in human monocytes are regulated differently. J Immunol 1991; 147:1530-1536.
25. Malyak M, Smith MF, Abel AA et al. Peripheral blood neutrophil production of IL-1ra and IL-1β. J Clin Immunol 1994; 14:20-30.
26. Jenkins JK, Malyak M, Arend WP. The effects of Interleukin-10 on Interleukin-1 receptor antagonist and Interleukin-1β production in human monocytes and neutrophils. Lymphokine Cytokine Res 1994; 13:47-54.
27. Kline JN, Fisher PA, Monick MM et al. Regulation of Interleukin-1 receptor antagonist by Th1 and Th2 cytokines. Am J Physiol 1995; 269:L92-98.
28. Muzio M, Re F, Sironi M et al. Interleukin-13 induces the production of Interleukin-1 receptor antagonist (IL-1ra) and the expression of the mRNA for the intracellular (keratinocyte) form of IL-1ra in human myelomonocytic cells. Blood 1994; 83:1738-1743.
29. Muzio M, Polentarutti N, Sironi M et al. Cloning and characterization of a new isoform of the interleukin-1 receptor antagonist. J Exp Med 1995; 182:623-628.
30. Kato K, Yokoi T, Takano N et al. Detection by in situ hybridization and phenotypic characterization of cell expressing IL-6 mRNA in human stimulated blood. J Immunol 1990; 144:1317-1322.
31. Wang P, Wu P, Anthes JC et al. Interleukin-10 inhibits Interleukin-8 production in human neutrophils. Blood 1994; 83:2678-2683.
32. Takeichi O, Saito I, Tsurumachi T et al. Human polymorphonuclear leukocytes derived from chronically inflamed tissue express inflammatory cytokines in vivo. Cell Immunol 1995; 156:296-309.

33. Fujishima S, Hoffman AR, Vu T et al. Regulation of neutrophil interleukin 8 gene expression and protein secretion by LPS, TNFα and IL-1β. J Cell Physiol 1993; 154:478-485.
34. Ember JA, Sanderson SD, Hugli TE et al. Induction of IL-8 synthesis from monocytes by human C5a anaphylotoxin. Am J Pathol 1994; 144:393-403.
35. Takahashi GW, Andrews DF, Lilly MB et al. Effect of GM-CSF and IL-3 on IL-8 production by human neutrophils and monocytes. Blood 1993; 81:357-364.
36. Konig B, Ceska M, Konig W. Effect of *Pseudomonas aeruginosa* on interleukin-8 release from human phagocytes. Int Arch Allergy Immunol 1995; 106:357-365.
37. Standiford TJ, Kunkel SL, Rolfe MW et al. Regulation of human alveolar macrophage and blood monocyte-derived IL-8 by PGE_2 and dexamethasone. Am J Respir Cell Mol Biol 1992; 6:75-81.
38. Wertheim WA, Kunkel SL, Standiford TJ et al. Regulation of neutrophil-derived IL-8: the role of prostaglandin E2, dexamethasone, and IL-4. J Immunol 1993; 151:2166-2175.
39. Strieter RM, Remick DG, Lynch III JP et al. Differential regulation of TNFα in human alveolar macrophage and peripheral blood monocytes: a cellular and molecular analysis. Am J Respir Cell Mol Biol 1989; 1:57-63.
40. Cavaillon JM, Marie C, Pitton C et al. Regulation of neutrophil derived IL-8 production by anti-inflammatory cytokines (IL-4, IL-10 and TGFβ). In: Faist E, ed. Proceedings 3rd International Congress on the Immune Consequences of Trauma Shock and Sepsis. Munich: Pabst Science Publishers, 1996, in press.
41. Haskill S, Peace A, Morris J et al. Identification of three related human GRO genes encoding cytokine functions. Proc Natl Acad Sci USA 1990; 87:7732-7736.
42. Ramenghi U, Ruggieri L, Dianzani I et al. Human peripheral blood granulocytes and myeloid leukemic cell lines express both transcripts encoding for stem cell factor. Stem Cells 1994; 12:521-526.
43. Roitt I, Brostoff J, Male D, eds. Immunology, 4th edition. Philadelphia, Mosby 1996:2-14.
44. Homburg CHE, Roos D. Apoptosis of neutrophils. Curr Op Hematol 1996; 3:94-99.
45. Colotta F, Re F, Polentarutti N, et al. Modulation of granulocyte survival and programmed cell death by cytokines and bacterial products. Blood 1992; 80:2012-2020.
46. Brach MA, deVos S, Gruss H et al. Prolongation of survival of human polymorphonuclear neutrophils by GM-CSF is caused by inhibition of programmed cell death. Blood 1992; 80:2920-2924.
47. Gosselin EJ, Wardwell K, Rigby WFC et al. Induction of MHC class II on human polymorphonuclear neutrophils by granulocyte/macrophage colony-stimulating factor, IFNγ and IL-3. J Immunol 1993; 151:1482-1490.

48. Mudzinski SP, Christian TP, Guo TL et al. Expression of HLA-DR on neutrophils from patients treated with granulocyte-macrophage colony stimulating factor for mobilization of stem cells. Blood 1995; 86: 2452-2453.
49. Smith WB, Guida L, Sun Q et al. Neutrophils activated by granulocyte-macrophage colony stimulating factor express receptors for interleukin-3 which mediate class II espression. Blood 1995; 86:3938-3944.

INDEX

Items in italics denote figures (f) or tables (t).

A

Actinomycin D (ACT D), 138
  effects on PMN cytokine expression, *139t, 140f,* 191

B

Bleomycin
  effect on kinetics of macrophage infiltration, 170
  PMN production of MCP-1 and, 97

C

C5a
  role and production during ischemia, 169
  PMN-derived IL-8, 183
CD30 ligand (CD30L)
  expression of CD30L mRNA by PMN, 98
Chemokines, 19, *20f*
Chronic granulomatous disease (CGD), 5
  IFN-γ and the treatment of, 128
Chronic inflammation
  cytokine gene expression in PMN derived from, 159
Colony-stimulating factors (CSFs), 9, 25
see also individual factors
Cycloheximide (CHX)
  effects on cytokine mRNA expression by PMN, 136, *137t*
Cytokines
see also individual cytokines
  defined, 9
  gene expression in stimulated PMN, 45, *46f*
  listed, by function and cell source, *10t*
Cytokine networks
  regulation by, 49-51, *50t*

D

Dexamethasone (DEX)
  modulatory effects on cytokine expression by PMN, 129

E

Ethanol
  effects on iNOS and TNFα expression in the lung, 168

F

formyl-Methionyl-Leucyl-Phenylalanine (fMLP)
  actions of, summarized, 183
  GROa and IL-8 release by PMN in response to, *92f*
  PMN production of IFNα or TGFβ in vitro, in response to, 60-62,*60f*

G

Gene regulation
  effects of metabolic inhibitors on, 136-139
Granulocyte-CSF (G-CSF), 26
  and PMN production of IFNa, 185
Granulocyte macrophage-CSF (GM-CSF), 25
Growth related gene product-alpha (GROα), 23, 91
  activities of, 93
  agonists of, 92-94, *94t*
  production and release by PMN, *92f,* 148, 190

H

Hepatitis C infection
  effects on TNFα production by IFNα, 160
HIV
  PMN cytokine production, 156

I

Inflammation, acute
  models involving PMN, 168
Integrins, 6
Interferon-alpha (IFNα), 11, 59-61, 185
  activity of, 11-12
  effects on TNFα production, 160
  IL-8 production by PMN and PBMC, effects on, 188, *189f*
Interferon-gamma, (IFNγ)
  activity of, 12
  and cytokine expression in PMN, 125-128
  and IL-8 secretion, 149
  co-inducer of IL-12 and IP-10, 184
Interferon inducible protein-10 (IP-10), 24
  production and release by PMN of, 97, 190
Interleukins,
  see individual interleukins
IL-1, 15, 169
  activities of, 16
  IL-1α, 66
  IL-1β, 66, 168
    cellular source of, 186
    list of stimuli inducing production of, 70, *71t*
    regulation of gene expression in PMN, 142, *143f*
  production by stimulated PMN, 67-68
  regulation of synthesis by human PMN, 69
  role in retinal ischemia, 170
  translational control in PMN, 146

IL-1 converting enzyme (ICE), 15-16
IL-1 receptor antagonist (IL-1ra), 70-77, 187
    agonists (listed), 76t
    bacterial infection, endogenous mediators of, 172
    extracellular release by PMN, effects of, 120f
    isoform of (icIL-1ra), 75
    LPS induction of, 72-73, 74f, 75
    mRNA stability in PMN, regulation of, 146
    regulation of secretion, 148
IL-2, 17
    stimulation of TNFα production by PMN, 64
Il-4, 17
    effect on IL-1β and TNFα mRNA expression, 115f, 115
    effects on PMN function by, 114-17, 115f, 116f
    translational control in PMN, 147
IL-6, 18-19, 187
    agonists of, 77f
    effects of LPS challenge, production of, 167
    kinetics of expression, 78
    mRNA expression 46f, 47f, 48f
IL-8, 19, 187
    agonists of, 83t, 84
    arthritis and, 171
    expression in PMN and monocytes, 48f, 80-87, 82f, 83t
        in CAPD patients, 158
        in periodontal infections, 159
    gene expression and protein secretion of, 81, 149
        in SLE patients, 158
        kinetics of expression, 82f
    inhibition of transcription in PMN by IL-10, 143, 143f
    PGE$_2$ and, 188
    production induced by chemoattractants, 81
    role during ischemia, 169
IL-10, 21
    and GROα inhibitory effects, 123-24
    effects on kinetics of mRNA expression of PMN-derived cytokines, 117, 119f, 185f
    effects on turnover rate of IL-1ra and IL-1β by, 144f, 146
    IL-1ra mRNA accumulation and, 187
    IL-8 production by PMN and PBMC, effects on, 189f
    modulatory effects on PMN cytokine release by, 117-125, 118f, 122
    regulation of the acute inflammatory response, 125
IL-12, 22, 87-91, 89f, 183
    agonists of, 91t
    genes and kinetics of expression in PMN of, 88, 190
    LPS and IFNγ induction of, 88
    therapeutic benefit in AIDS, 22
IL-13, 17
    IL-1ra production and, 187
Interleukin system, 15-22
see also individual interleukins
Ischemia
    roles of C5a and IL-8 during, 169

L
Lipopolysaccharide (LPS)
    effects on PMN cytokine expression in vivo, 163-168
    -elicted acute inflammation, lung model of, 165-166

M
Macrophage inflammatory protein (MIP)-1α and -1β, 23-24
    agonists of, 96t
    effects on gene expression by IL-10, 145
    regulation of expression, 95
    transcriptional regulation in PMN of, 143-145
    translational control in PMN, 147
Melanoma growth stimulatory activity (MGSA), 23
see GROα
Microcrystals,
    CPPD (calcium pyrophosphate dihydrate), 147
    MSU (monosodium urate monohydrate), 147
    effects on cytokine expression by PMN of, 128-129
Monocyte chemotactic proteins (MCP-1, MCP-2, and MCP-3), 24, 96-97
    agonists of, 96t
    mRNA expression and lung injury, 170
Myeloperoxidase (MPO)
    -containing granules in PMN, 4
Myocardial infarct, rabbit model, 164t, 183
    ischemia
        C5a and IL-8, role during, 169

N
NADPH oxidase, 4
    activity of, 5
Neutrophils (polymorphonuclear leukocytes, PMN)
    antimicrobial mechanisms of, 4
    antitumor activity, in vivo and, 176
    culture conditions for, 40
    cytokine production of, in-vitro, 40-41, 41f, 43, 44f
        intracellular control of, 190-191
        in-vivo, 51-53, 53f, 164t
            distinct patterns of release, elicited by different stimuli, 181-184
        microcrystals and, 128-129
        molecular regulation of, 141t
        stimuli of, 41, 42t
    differentiation, 1-2
    functional responses to agonists of, 3t
    modulation of function by IL-4, 114-17, 115f, 116f
    receptors of, 2
    role in acute inflammation, 5-6
    translational control for IL-1, 146
    transcriptional regulation in, 139-146

P
Polymorphonuclear leukocyte (PMN)
see Neutrophil
Prostaglandin E$_2$ (PGE$_2$)
    and mRNA expression of LPS-induced PMNs, 129
    PMN-derived IL-8, 188

R

RANTES, 23, 97, 165
Reactive oxygen intermediates (ROI), 4
Respiratory syncytial virus (RSV)
   and IL-8 mRNA levels, 84
Rheumatoid arthritis
   PMN-derived cytokines and, 156-158
   role of GROα in SF, 157

S

Saccharomyces cerevisiae
   opsonized with IgG (Y-IgG), as a stimulus of PMN-cytokine production and release, summarized, 182
   effect on PMN TNFα release, 63, 66*t*
   GROα release in response to, 93
Selectin family, 6
Sepsis syndrome (sepsis-PMN), 68, 155
   mechanism for tolerance to endotoxin, 155-156
Shigellosis
   cytokine production, 159-160
Sodium nitroprussiate (SNP)
   effects on LPS-dependent TNFα production by PMN, 186
Staurosporine
   induction of IL-8 mRNA, 86
Stem cell factor (SCF), 25
   expression in PMN, 190

Substance P
   stimulation of IL-8 release by, 86
Sulfatides, 65
   induction of IL-8 mRNA, 81
Systemic lupus erythematosus (SLE)
   IL-8 production by PMN, 158

T

T cell subsets, 174
   activation and outcome of yeast infection, 174
Transforming growth factor-β, (TGFβ), 186
   actions of, 13
   expression and secretion by neutrophils, 61-62
   localization in developing endochondral bone, regulation of ossification, 177
Tumor necrosis factor-alpha (TNFα), 14, 186
   and Crohn's disease, 160
   during acute inflammation, production of, 165
   expression of TNFα mRNA by neutrophils, 62
   stimulation of IL-8 production by, 82*f*
   stimuli of, 63, 65, 66*t*, 186
   translational regulation of, 150
   and tumor regression, 176

Z

Zymosan
   stimulation of, 85